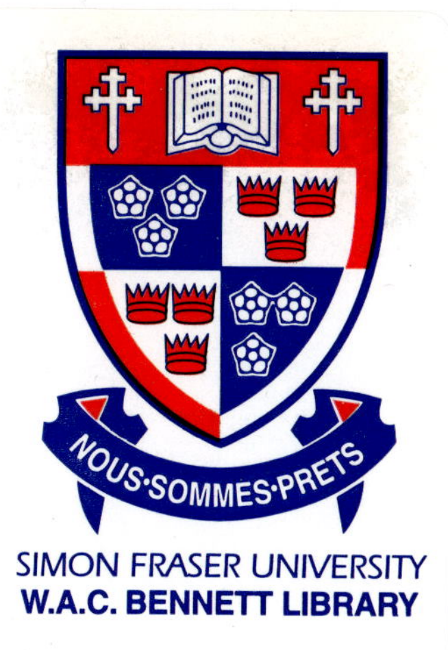
NOUS·SOMMES·PRETS
SIMON FRASER UNIVERSITY
W.A.C. BENNETT LIBRARY

D0151599

RENEWING URBAN COMMUNITIES

Renewing Urban Communities

Environment, Citizenship and Sustainability in Ireland

Edited by

NIAMH MOORE and MARK SCOTT
University College Dublin, Ireland

ASHGATE

Published by
Ashgate Publishing Limited
Gower House
Croft Road
Aldershot
Hampshire GU11 3HR
England

Ashgate Publishing Company
Suite 420
101 Cherry Street
Burlington, VT 05401-4405
USA

Ashgate website: http://www.ashgate.com

British Library Cataloguing in Publication Data
Renewing urban communities : environment, citizenship and sustainability in Ireland. - (Urban planning and environment)
1. Community development, Urban - Ireland 2. Sustainable development, - Ireland 3. Cities and towns - Ireland - Growth 4. Ireland - Social conditions - 21st century
I. Moore, Niamh II. Scott, Mark
307.1'416'09415

Library of Congress Control Number: 2005923317

ISBN 0 7546 4083 3

Printed and bound in Great Britain by MPG Books Ltd, Bodmin, Cornwall

Contents

List of Figures *vii*
List of Tables *ix*
List of Contributors *x*
Preface *xv*
Acknowledgements *xvi*

1 Introduction: The Geographical and Policy Context
Niamh Moore and Mark Scott 1

PART I: SUSTAINABLE URBAN ENVIRONMENTS

2 Re-greening Brownfields: Land Recycling and Urban Sustainability
Niamh Moore 29

3 Higher Density or Open Space: The Future Role of Green Space
Craig Bullock 49

4 Suburbanising Dublin: Out of an Overcrowded Frying Pan into a Fire of Unsustainability?
Andrew MacLaran 60

5 Urban Form and Reducing the Demand for Car Travel: Towards an Integrated Policy Agenda for the Belfast Metropolitan Area?
Malachy McEldowney, Tim Ryley, Mark Scott and Austin Smyth 75

6 Improving Energy Efficiency in Urban Areas
J. Peter Clinch 94

7 From Barricades to Back Gardens: Cross-Border Urban Expansion from the City of Derry into Co. Donegal
Chris Paris 114

8 Urban-Generated Rural Housing and Evidence of Counterurbanisation in the Dublin City-Region
Menelaos Gkartzios and Mark Scott 132

PART II: SUSTAINABLE COMMUNITIES

9 'We're Too Busy for That Kind of Stuff': Progress Towards Local Sustainable Development in Ireland
Geraint Ellis, Brian Motherway and William J.V. Neill 159

10 Social and Ethnic Segregation and the Urban Agenda
Brendan Murtagh 179

11 The State and Civil Society in Urban Regeneration: Negotiating Sustainable Participation in Belfast and Dublin
Jenny Muir 197

12 Active Citizenship: Resident Associations, Social Capital and Collective Action
Paula Russell, Mark Scott and Declan Redmond 213

13 Housing Policy, Homeownership and the Provision of Affordable Housing
Declan Redmond and Gillian Kernan 235

PART III: CONCLUSION

14 Towards a Sustainable Future for Irish Towns and Cities
Niamh Moore and Mark Scott 253

Index *263*

List of Figures

1.1	Comparative city-size in the Republic of Ireland	2
1.2	Regional development strategy for Northern Ireland	17
1.3	Spatial development strategy for the Republic of Ireland	18
1.4	Revised spatial planning guidelines for the GDA	19
2.1	Influential factors in brownfield redevelopment	31
2.2	Open cast mining in Northern Bohemia	32
2.3	Uranium mine at Jachymov, Bohemia	33
2.4	LeBreton Flats in relation to the city centre	34
2.5	The Grand Canal Dock, 1876	41
2.6	Grand Canal Dock prior to redevelopment	43
2.7	Tenants at the Grand Canal Harbour redevelopment	45
2.8	Spin-off developments from Grand Canal Harbour	46
4.1	Dublin suburban locations	63
4.2	Dublin postal districts	65
4.3	Atrium I and II, Sandyford, Dublin 18	67
4.4	Liberty Hall, Dublin 1 (1965)	68
4.5	Park West Business Park, Dublin 22	68
5.1	Regional development strategy for Northern Ireland	80
5.2	Development context for the Belfast Metropolitan Area	81
5.3	Connections between land-use and transportation policy	82
6.1	Indices of GDP per capita and energy consumption by sector	95
6.2	Indices of GDP, vehicle numbers and selected air emissions, 1990-2002	96
6.3	'Business as usual' greenhouse gas emissions, 1990-2010 and associated target for 2012	97
6.4	Selected emissions, 1990-2002 and associated targets for 2010	98
6.5	Vehicle numbers and related CO_2 emissions, 1990-1998 and forecasts to 2010	99
6.6	Energy consumption by transport, 1990-2010	100
6.7	Domestic coal consumption in Ireland in 1000s TOE	103
7.1	Religious distribution in Derry City	116
7.2	The Derry city-region	127
8.1	Preferences for rural or suburban lifestyle in Ireland	137
8.2	Spatial strategy for the GDA	139

8.3	The Dublin city-region	142
8.4	Percentage change in the population of each county in Leinster at each census since 1979	143
8.5	Percentage change in the population of electoral divisions in Leinster, 1996-2002	143
8.6	County Louth population changes, 1996-2002	144
8.7	County Meath population changes, 1996-2002	144
8.8	County Kildare population changes, 1996-2002	146
8.9	County Wicklow population changes, 1996-2002	146
8.10	Private house completions in Leinster, 1996-2003	147
8.11	Percentage of owner-occupied private dwellings in each county in Leinster, 2002	148
8.12	Average new house prices (excluding apartments)	149
8.13	Average existing house prices (excluding apartments)	149
8.14	Percentage of people who travel to work, school or college by car (driving or passengers)	151
8.15	Percentage of people who travel 15+ miles from home to work, school or college in Leinster	152
9.1	Percentage of councils actively engaged in LA21 policies	166
9.2	Dimensions of LA21 strategies engaged in by local authorities	166
9.3	Relevance of LA21 to the promotion of sustainable development	169
10.1	Key locations in the Belfast Metropolitan Area	183
10.2	Interface at Cluan Place, East Belfast	184
10.3	Interface in Manor Street, Urban II area	188
13.1	New housing completions: State	237
13.2	New house prices in the State and Dublin	238
13.3	Housing costs as a proportion of household expenditure (%)	241
13.4	Ratio of house prices to average industrial wages (%)	242

List of Tables

1.1	Population change in Ireland, 1911-1966	3
3.1	Attribute coefficients and marginal economic values	56
4.1	Location of the modern office stock, December 2003	69
4.2	Daily vehicle-based movement, Paris 1998	72
5.1	A summary of the household survey sample	85
6.1	Breakdown of greenhouse gas emissions by sector, 1990 and 2010	101
7.1	Population change in Derry, 1991-2001	116
7.2	Regional population change in Northern Ireland and Republic of Ireland, 1991-2001/2	120
7.3	Household tenure in the Republic of Ireland and Northern Ireland, 1991-2001/2	122
7.4	Population change in County Donegal, 1996-2002	126
7.5	Population change in Derry border zone, County Donegal, 1996-2002	127
8.1	Choice of areas in the short, medium and long-term	137
8.2	Percentage changes of rural house prices in Dublin and the rest of Leinster, 2001-2003	150
10.1	Community interaction and mobility in East Belfast	185
10.2	A SWOT analysis of North Belfast	190
12.1	Profile of chosen areas	220
13.1	Dwellings by type of tenure in Ireland	236
13.2	House prices and Consumer Price Index	240
13.3	Affordable house completions: State	246
14.1	Indicators of quality of life applied to Ireland	255

List of Contributors

Craig Bullock is associated with the Department of Planning and Environmental Policy at University College Dublin and also manages Optimize, a consultancy specialising in environmental and socio-economic analysis and evaluation. His particular interests include environmental policy, environmental valuation methods, impact assessment, urban planning/quality of life, and rural development. He has ten years experience of working in applied research and additional experience of the private sector and of agricultural development overseas.

J. Peter Clinch is concurrently Jean Monnet Professor of European Environmental Policy and Professor of Regional and Urban Planning at University College Dublin. He is also Head of the Department of Planning and Environmental Policy. He has recently held visiting positions at the University of California, Berkeley and the University of California, San Diego. He is author of over 80 publications including four recent books and in 2003 was appointed by the EU to the Jean Monnet Chair of European Environmental Policy in recognition of his research and scholarship in this field.

Geraint Ellis is Senior Lecturer in the School of Environmental Planning and Deputy Director of the Centre for Sustainability and Environmental Governance at Queen's University, Belfast. He has previously worked in the community and environmental sectors in London and as a development worker in Southern Africa. He has published a range of academic research on land use planning and sustainable development, with key interests in equality and environmental justice. Together with Brian Motherway and Bill Neill, he has recently completed a major report on local sustainable development in Ireland for the Centre for Cross Border Studies. He is a member of the Department of the Environment (NI)s advisory group on the Northern Ireland Sustainable Development Strategy and a director of a number of voluntary organisations that includes Community Technical Aid, Belfast Healthy Cities and Sustainable NI.

Menelaos Gkartzios is a graduate of the Agricultural University of Athens and University College Dublin with degrees in Agricultural Economics and Rural Development and Environmental Resource Management and is currently a Researcher in the Urban Institute Ireland and Department of Planning and Environmental Policy, University College Dublin. His research interests include rural housing and consumer preferences, counterurbanisation trends, rural development and urban/rural relationships. Menelaos is currently undertaking research on metropolitan decentralisation in the Dublin city-region, which is funded by the Higher Education Authority through the Urban Institute Ireland.

Gillian Kernan graduated from University College Dublin in 2002 with a B.A. (Hons) in Economics and Geography. In 2003 she received her MSc in Economics from UCD. She is now a PhD student in the Department of Planning and Environmental Policy, UCD, where she has undertaken research evaluating the provision of affordable housing in Ireland. This research is being carried out with Dr. Declan Redmond and is funded by the Combat Poverty Agency. She is now in the process of completing a PhD in the area of house price affordability.

Andrew MacLaran is a Senior Lecturer in Geography and Director of the Centre for Urban & Regional Studies, Trinity College Dublin. His research interests focus on the institutional forces shaping the city, notably the impact on urban communities and the built environment of the property development sector and the manner in which urban planning has attempted to cope with and influence the outcomes of the development process. He was awarded the Manning Robertson Prize in 1995 by the Irish Branch of the Royal Town Planning Institute, for the contribution made by *Dublin: the shaping of a capital* (Belhaven-Wiley, 1993) to a better understanding of Irish urban planning. Recently, Andrew has edited *Making Space: property development and urban planning* (Arnold, 2003), which reviews the private-sector forces responsible for urban development, together with the systems of urban planning put in place to influence, guide and manipulate the outcomes. He is currently joint editor of the *Journal of Irish Urban Studies*.

Malachy McEldowney is Professor of Town and Country Planning at Queen's University Belfast and was Head of the School of Environmental Planning from 1993-2002. He is an architect planner, having worked in Leicester City Planning Department for several years before entering academic life. In 1984 he was Visiting Professor in the Graduate Program in Urban Planning in the University of Kansas, USA, and in 1988 he was Visiting Professor to the School of Architecture and Urbanism in the University of Niteroi, Rio de Janeiro, Brazil. In recent years his research interests have focused on strategic spatial planning and transportation, having been a member of the Research Consortium which carried out the public consultation for the Regional Development Strategy for Northern Ireland and the Belfast Metropolitan Area Plan, and having been a facilitator for the consultation exercises on the Regional Transportation Strategy and the Belfast Metropolitan Transport Strategy.

Niamh Moore is a Lecturer in the Department of Geography, University College Dublin. Her research interests are in the regeneration of former industrial sites, focusing on issues of industrial heritage, land use change, sustainability and social inclusion. She has previously been an Urban Institute Ireland scholar and a Government of Ireland Post-doctoral Fellow. As well as teaching undergraduates, she lectures on a European urban historical geography Masters course to architecture, geography, urban design and European studies students. Niamh has published in Irish and international journals and is the author of the forthcoming books *Reinventing Dublin Docklands: People, politics and place* (Four Courts Press, 2005) and with Dr Yvonne Whelan (University of Bristol), *Heritage, memory and the politics of space: New perspectives on the cultural landscape* (Ashgate, forthcoming). Niamh is a Research Associate of the Urban Institute Ireland and is Co-convenor of its Housing and Sustainable Communities Research Cluster.

Brian Motherway is a Director of the consulting firm Motherway Begley, working in the area of social and policy aspects of environmental issues. Areas of interest include public participation in decision making, environmental attitudes and behaviour and sustainable development policy. His PhD is from Trinity College Dublin, for research into the sociological aspects of public participation in environmental debates in Ireland. He also holds Bachelors and Masters degrees in chemical engineering.

Jenny Muir is a Research Fellow at the Institute of Governance, Public Policy and Social Research at Queen's University, Belfast. Jenny has a PhD from the University of Ulster, after which she worked as a postdoctoral research fellow at the Urban Institute of Ireland, University College Dublin and as a visiting research fellow at Flinders University, Adelaide, Australia. Before moving to Northern Ireland, Jenny worked for several London boroughs in housing policy and regeneration roles. Her research interests include: urban regeneration policy and practice; public participation in urban regeneration programmes; governance issues and public policy; and tenant participation in the planning and management of social housing.

Brendan Murtagh is a Reader in the School of Environmental Planning at Queens University Belfast. He has researched and written widely on urban regeneration, conflict and community participation. His book on the *Politics of Territory* (Macmillan, 2002) addressed the complex challenges of planning in a divided society and drew on both local and international examples of best practice. He has recently co-authored a book on rural planning and community dialogue (with M. Murray, *Equity, Diversity and Interdependence*, Ashgate, 2004) and on *Urban Regeneration in Divided Cities* (with P. Shirlow, forthcoming, Pluto Press).

William J.V. Neill is Director of postgraduate planning studies at Queen's University Belfast. He graduated in planning from the University of Michigan and worked for many years in economic development for the State of Michigan. His most recent books are *Urban Planning and Cultural Inclusion* (ed. with Hanns-Uve Schwedler, Palgrave, Macmillan 2001) and *Urban Planning and Cultural Identity* (Routledge, 2004).

Chris Paris is Professor of Housing Studies at the University of Ulster, and holds visiting chairs in planning and social science at the University of Hong Kong and RMIT University in Melbourne. Chris has over thirty years experience of research and teaching in universities in the UK, Australia, Ireland and Hong Kong. He has published extensively in scholarly and professional journals and written books on housing and planning in the UK, Ireland and Australia. His most recent book *Housing in Northern Ireland – and comparisons with the Republic of Ireland* was published in the CIH 'Policy and Practice' Series. He recently completed research on demographic change and housing need and is currently working on demography and housing, housing markets and urban-regional change in Ireland, and comparative aspects of housing development.

Declan Redmond is a Lecturer in the Department of Planning and Environmental Policy at University College Dublin. He is Programme Director of the Master of Regional and Urban Planning degree, accredited by the Irish Planning Institute and the Royal Town Planning Institute. His principal research interests revolve around housing and planning, focusing in particular on changing urban governance as well as the provision and management of social and affordable housing. He has undertaken doctoral work on regeneration and community participation and is currently involved in research projects investigating housing affordability, the role of private residents associations in the planning system, the participation of disadvantaged communities in the planning system and the role of entrepreneurial governance in urban change.

Paula Russell is a Lecturer in the Department of Planning and Environmental Policy, University College Dublin. Her main research interests lie in the areas of urban regeneration (including the community impacts of urban regeneration and the place of regeneration in the wider process of urban governance), participatory planning and the interface between planning and civil society. Her current research projects include, a study on residents' associations, neighbourhood and community development and a study of the potential of market based instruments in creating more sustainable settlement patterns.

Tim Ryley is a Lecturer in Transport Studies at Loughborough University. Prior to Loughborough University, he had ten years transport research experience gained at the Transport Research Laboratory, the University of Ulster and most recently Napier University. Tim Ryley's research interests include the relationship between urban form and travel behaviour, the acceptability of road pricing, the promotion of non-motorised modes and the influence of the media upon transport projects. Tim Ryley was employed as a researcher on two successive Sustainable Cities projects between 1998 and 2002, funded by the Engineering and Physical Sciences Research Council (EPSRC). The projects examined the relationships between transport, planning and housing for the cities of Belfast and Edinburgh, and involved the development of a range of transport models, questionnaire surveys and stated preference experiments.

Mark Scott is a Lecturer in the Department of Planning and Environmental Policy, University College Dublin. He is a graduate of Queen's University Belfast with a PhD in Environmental Planning, and has previously held research posts in the University of Ulster and Queen's University Belfast. Mark's research interests revolve around the interface between spatial planning, local governance and sustainable development. Ongoing research projects focus on residential change in the inner city and the rural-urban fringe, the role of civil society in managing environmental change in both urban and rural communities, and spatial policy in remote rural areas. This research is funded by the Irish Research Council for Humanities and Social Sciences, the Royal Irish Academy and the Environmental Protection Agency. Mark is currently Director of the PhD Programme in Planning and Environmental Policy in UCD and is also a Research Associate of the Urban Institute Ireland and Co-Convenor of its Housing and Sustainable Communities Research Cluster.

Austin Smyth is Professor of Transport Economics in the Transport Research Institute, Napier University, and also Director of the Transport Research Institute Northern Ireland Centre at Queen's University Belfast and Director General of the National Institute for Transport and Logistics at the Dublin Institute of Technology. Austin has previous experience as Director Designate with Ove Arup and Partners (London) and as Co-Director of the University of Ulster's Transport and Road Assessment Centre (TRAC) and Chair of the University of Ulster's School of the Built Environment Professorate. His research interests include transport planning and operations analysis, urban transportation modelling, stated preference and attitudinal research techniques, and appraisal, evaluation and assessment procedures, on all of which he has published widely.

Preface

Ireland is now an urban society, and both parts of the island have experienced rapid urban-generated growth and new patterns of development in recent years. Emerging trends have included: suburban growth and edge of city development; an increase in the spatial separation of home and workplace as the commuter belt for urban areas enlarges; counterurbanisation; increased traffic congestion; housing affordability issues; and a policy response that increasingly favours sustainable urban management such as increasing urban densities and integrating land-use and transport planning. Related to these trends have been changing approaches to urban governance and issues of active citizenship and social exclusion. In response to this emerging restructured geography of urban Ireland, a range of strategic planning initiatives has recently been undertaken. These include the Regional Development Strategy in Northern Ireland, and the Strategic Planning Guidelines for the Greater Dublin Area and National Spatial Strategy and in the Republic of Ireland. But even with all of these in place, there appears to be an ever-increasing sense that growth and change is still taking place in a haphazard and uncontrolled fashion. It is perhaps then timely to take stock, to see where Ireland is at and whether the policy instruments currently being exhorted and lauded can actually achieve their objectives. This book investigates the tension that exists between sustainable urban development values and rhetoric and the emerging geography of urban Ireland, influenced by consumer and lifestyle choices.

Driving this project has been the increased focus on inter-disciplinarity and policy relevance within the academy and indeed within our own departments. Increasingly, it is understood that solutions to the kind of problems addressed in this book can only be found by combining the resources of a number of different academic disciplines. In that context, this book is the product of the Housing and Sustainable Communities Research Cluster at the Urban Institute Ireland (UII). Created under the leadership of University College Dublin in partnership with Trinity College Dublin, and funded by the Irish Higher Education Authority's Programme for Research in Third Level Education (PRTLI), Urban Institute Ireland aims to maximise the research potential of many of Ireland's leading academics by adopting a creative interdisciplinary and collaborative approach. This book closely follows the ethos of the UII by examining the issue of urban sustainability from a cross-disciplinary and cross-border perspective. It is our hope that this book will succeed in generating a debate, not only about the provisions of current Irish urban policy, but also about its future direction in a dramatically transformed, and transforming, economic and political environment.

Acknowledgements

The Editors wish to acknowledge and thank a range of people involved in the production of this book. Funding for this publication was generously provided by the Urban Institute Ireland at University College Dublin, through the auspices of the Higher Education Authority's Programme for Research in Third Level Education (PRTLI). In this regard, we would like to take this opportunity to thank John Yarwood (former Director of UII), Professor Frank Convery (present Director) and Seona Meharg (Development Manager) for their continuous support for this project and throughout the preparation of the book. Thanks are due to the Ordnance Survey Ireland for permission to reproduce some of their maps in this publication (Permit no: 8015).

We would also wish to thank Professor Gert de Roo, co-editor of the *Urban Planning and Environment Series* with Ashgate for his helpful comments and advice, and for their administrative and technical input and support, a big thank you to Val Rose, Emily Poulton, Maureen Mansell-Ward and Carolyn Court of Ashgate Publishing. We are also grateful to Stephen Hannon, Department of Geography, University College Dublin for the preparation of the majority of the maps in this volume, and to Carole Devaney for undertaking the onerous job of completing the indexing.

We are very thankful to our many colleagues who contributed chapters to this book for their collaboration and timely input. Finally, we are indebted to the large number of anonymous referees from numerous disciplines and institutions on the island of Ireland for their helpful feedback and comments on each chapter, which have helped to bring this book to its conclusion.

Niamh Moore and Mark Scott
April 2005

Chapter 1

Introduction: The Geographical and Policy Context

Niamh Moore and Mark Scott

Introduction

> Sustainable human life on this globe cannot be achieved without sustainable local communities. Local government is close to where environmental problems are perceived and closest to the citizens and shares responsibility with governments at all levels for the well-being of humankind and nature. Therefore, cities and towns are key players in the process of changing lifestyles, production, consumption and spatial patterns (Charter of European Towns and Cities Towards Sustainability).

Ireland is now an urban society, and both parts of the island have experienced rapid urban-generated growth and new patterns of development in recent years. Emerging trends have included: suburban growth and edge of city development; an increase in the spatial separation of home and workplace as the commuter belt for urban areas enlarges; counterurbanisation; increased traffic congestion; housing affordability issues; and a policy response that increasingly favours sustainable urban management such as increasing urban densities and integrating land-use and transport planning. Related to these trends have been changing approaches to urban governance and issues of active citizenship and social exclusion. In response to this emerging restructured geography of urban Ireland, a range of strategic spatial planning initiatives have recently been undertaken. These include the Regional Development Strategy in Northern Ireland, and the Strategic Planning Guidelines for the Greater Dublin Area and National Spatial Strategy in the Republic of Ireland.

This chapter provides an overview of the dynamics of urban change in Ireland, particularly during the 1990s and the experience of rapid economic growth. It will situate contemporary urban development within the extensive literature on sustainable cities and the restructuring of urban governance, while also examining recent policy initiatives at a European, national, regional and city-region scale.

While Dublin has always dominated the urban pattern of Ireland, the other cities in the Republic of Ireland remain relatively small in population as the rank-size graph shows (Figure 1.1). The larger cities are coastal in location, and the

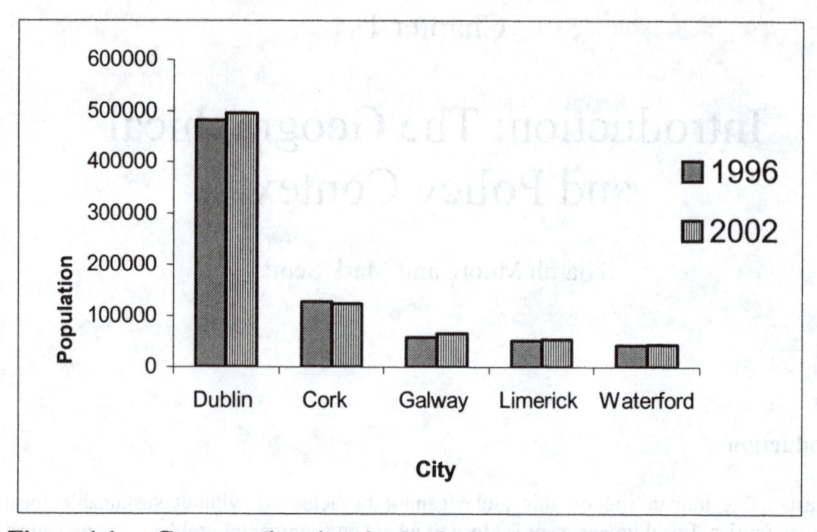

Figure 1.1 Comparative city-size in the Republic of Ireland

centre of the island is relatively unpopulated. There has long been a concern with the imbalance between the scale of the Dublin region and the rest of the country in terms of population and prosperity. The Greater Dublin Area[1] has 39 per cent of national population, and official policy is concerned to ensure that it does not exceed that proportion. It is recognised as the economic engine of the country, however, and some people argue that to 'cap' its growth will retard the economic development of the state, especially as the global location market becomes more fiercely competitive.

This is one of the major themes of the National Spatial Strategy 2002-2020, and remains a major matter of controversy. There is a historic echo here, as the 'Pale', a defensive fortification around the Dublin region built in mediaeval times to keep the countrymen at bay, is traceable in the emergent patterns of urban culture and structure today.

After the appalling privations of the nineteenth century, although the population of Northern Ireland continued to grow, that of the Republic of Ireland continued to lose population and jobs between the First World War and the early 1960s (Table 1.1). Since the adoption of new economic policy signalled by the First Programme for Economic Expansion of 1958, the economy has developed in an astonishing way, and at the present time only Luxembourg has a higher GDP per capita in Europe, and in the world as a whole, only the USA exceeds it. The birth rate is twice the death rate, which is unique in Europe (except for Turkey). The growth of the Greater Dublin Area has been driven by these phenomena, and arguably the rate of change has outstripped the capacity of both societal and public administrative resources to cope with it. One major development issue here has been density.

Table 1.1 Population change in Ireland, 1911-1966

Year	*Republic of Ireland*	*Northern Ireland*	*Total*
1911	3,139,688	1,250,531	4,390,219
1926	2,971,992	1,256,561	4,228,553
1936	2,968,420	1,279,745	4,248,165
1951	2,960,593	1,370,921	4,331,514
1961	2,818,341	1,425,042	4,243,383

Source: Census of Population

Up to the 1930s, Dublin was a 'compact' city, with the average density in inner Dublin over 100 persons per hectare (pph), rising to 450 persons per hectare in the central wards, where there was bad overcrowding. However, by 1961, there were only 160,000 persons living in the inner city, 100,000 fewer than in 1936 due in part to large-scale slum clearance and extensive suburban expansion. Plans typically set density guidelines around 25 dwellings to the hectare (or ten per acre.) The average annual expansion from 1936 to 1972 was less than 300 hectares, but this grew to an annual rate of 450 hectares between 1973 and 1988. The rate of expansion was much greater than the growth of population and consequently, gross density reduced from over 85 pph in 1936 to 50 in 1988. The population within 20 miles of the city centre doubled to 1.2 million persons, most being accommodated in two-storey single-family homes with large gardens and reliant on private transportation. Following the suburbanisation of population, the suburbanisation of both retailing and office space occurred throughout the 1980s and 1990s, further complicating the urban system and making it increasingly difficult to consider how best to achieve a frequent and viable public transportation network. The spreading of Dublin's physical form is thus a long term trend, and today the region is characterised by low densities, and long commuting distances. Recent research by Carroll (2004, p. 113) has illustrated how unsustainable the city is, stating that "the land required to support the lifestyle and activities of Dublin residents is 2.34 times the size of Ireland … if each person on earth lived like a Dubliner, almost three earths would be required to provide natural resources".

Broadening the debate, we should mention two other key policy issues of a regional scale. The first is the emergence of an 'edge city' related to the orbital M50 motorway, reflecting in particular the development of 2.3 million sq. m. of offices from 1996 to 2002, of which 67 per cent was built in the south-east suburbs near the M50. This has been driven by the withdrawal of much central government financial support to local authorities, requiring the latter to compete with each other to attract office and retail parks and so to build up their tax bases.

The second issue is 'counterurbanism', that is, the tendency of houses (one third of permits) to be built as 'one-off' houses in rural areas to accommodate people who work in the metropolis. McDonald (2002, p. 317) has questioned this, arguing that "the development of 'sustainable communities' cannot be advanced by allowing Dublin, willy-nilly, to sprawl all over Leinster. Yet who among the authorities will shout stop? Who will admit that what's going on is profoundly unsustainable? ...Whatever about all the official guff about 'sustainable development', nobody can ignore facts on the ground. And those facts include the staggering statistic that 18,000 of the record 50,000 new homes completed last year consisted of one-off houses in the countryside. Bungalow Blitz and Mansion Mania rule in the Celtic Tiger's lair".

Whereas urban change in the Republic of Ireland has been largely driven by rapid economic growth, in Northern Ireland urban change has primarily resulted from political factors. Belfast, the capital city of Northern Ireland, is infamous across the world for being synonymous with political violence, and for the last three decades the city has suffered from the economic and political consequences of an intense sectarian conflict that has shaped the city's management (Ellis and McKay, 2000). This political context has reinforced deep-seated structural problems in the Northern Ireland economy. Historically, the north-east of Ireland was the most industrialised part of the island and a century ago, Belfast was the industrial heartland with a global status as a production site (Gaffikin and Morrissey, 1999a) – based on shipbuilding, engineering and linen. However, the narrowness of its industrial base made Northern Ireland vulnerable to the structural changes of the post-war economy, and since the 1960s huge numbers of jobs have been lost in all manufacturing sectors.

In population terms, the Belfast Metropolitan Area (BMA)[2] is the largest urban centre in the region with an estimated population of 650,000 people (DOE, 2001), which is almost 40 per cent of the total population of Northern Ireland. Over one million people live within a 30-mile radius of Belfast City Centre. Since the 1960s, urban development has been characterised by the emergence of a low-density urban form and by population loss in the central city, as the BMA has experienced massive out migration from the urban core, common to most UK cities, although exacerbated by the 'Troubles' (Cooper et al., 2001). Between 1971 and 1981, there was a period of intense urban decentralisation of Belfast when the population of the urban core fell from 416,700 to 314,300 (Ellis and McKay, 2000) and continued to decline throughout the 1980s. By 1991 it had fallen to 279,200, as people continued to move out of the city into the contiguous suburbs and dormitory towns and villages. Ellis and McKay (2000) suggest that although a significant degree of population movement occurred as a result of the violent political conflict, there were a number of other reasons for this population shift, for example:

- Comprehensive redevelopment of the inner city;
- Growth of private housing;

- Decline in mean household size;
- Rising car ownership resulting in increased commuting;
- Regional policy, encouraging out migration from Belfast as a policy of demagnetising Belfast was pursued in the 1970s with a more diffuse settlement strategy.

A central feature of the social geography of Belfast over the past thirty years has been the high degree of residential segregation with the principal division being by religion (Protestant and Catholic). There is a long history of religious segregation in Belfast, dating back at least 200 years (Boal, 1995). However, ethno-religious spatial segregation has intensified during the political violence of the contemporary 'Troubles'. As recorded by McPeake (1998), throughout the past thirty years segregated space in Belfast has been purified and consolidated such that the ethnic boundaries have become more clearly defined. McPeake outlines that by 1991, of the 117 electoral wards that make up the city, 62 were more than 90 per cent one religion or the other; and clearly "as far as housing is concerned, Belfast can justifiably be described as a divided city" (1998, p. 529). Moreover, even in the more prosperous 'middle-class' residential areas which exhibit less segregation, social networks and social capital are still assembled along familiar ethno-religious lines, often built around church and schools which helps to reproduce asymmetrical patterns of behaviour, contact and networks (Murtagh, 2000).

Fresh impetus for change has arisen from the search for a political settlement in Northern Ireland. The 'peace process' began in the early 1990s with the announcements of a cessation of violence by republican and loyalist paramilitaries. This culminated with the Good Friday Peace Agreement signed by the British and Irish Governments and the main Northern Ireland political parties on 19 April 1998, and subsequently endorsed in referenda in both Northern Ireland and the Republic of Ireland. The Agreement established new political structures in Northern Ireland, including a new Assembly and a cross-party Executive, however, at the time of writing these have been suspended as the peace process has repeatedly stalled. These ongoing political negotiations notwithstanding, Northern Ireland appears to have entered a new era of relative stability and non-violence, providing an opportunity to reshape Northern Ireland society. The impact on urban life of this new political context has been outlined in a recent special report in *The Economist* (2004):

> Belfast, like Dublin, has been transformed over the past 15 years. It used to be a bleak, partly bombed-out shell of a Victorian city, festooned with barbed wire and full of nervous soldiers in armoured vehicles. Now its centre is a bustling, lively place full of shops, offices, restaurants and theatres … Overall Northern Ireland has changed sharply for the better.

Since 1991, there has been a small growth in the population within the Belfast urban core following decades of decline (DOE, 2001). This has included widespread apartment developments in the city centre, providing for the first time in over thirty years opportunities for living in an area that during the 1970s and 1980s was essentially cordoned off outside working hours as a security measure. However, regardless of the peace process, residential segregation remains an enduring feature of the city's geography.

Sustainable Development and Urban Management

In a recent paper, Briassoulis (1999) highlighted how sustainable development is now commonly cited as the ultimate urban planning goal, although what it means is not usually specified exactly nor how it is to be achieved. In general, participants in urban management processes agree that sustainability is concerned with the simultaneous satisfaction of three objectives – environmental protection, social equity and economic development (Lindsey, 2003); in other words, creating a positive-sum strategy embracing these three policy goals (Albrechts et al., 2003). In this regard, urban management tools and policies are a "key arena within which economic, social and environmental issues come together with respect to the spatial dimensions of management of environmental change" (Healy and Shaw, 1993, p. 770).

The most widely known and used definition of sustainable development is that provided by the World Commission on Environment and Development which suggests that sustainable development is "development which meets the needs of the present without compromising the ability of future generations to meet their own needs" (1988, p. 8). The concept of sustainable development has emerged as the central and unifying theme of a new environmental agenda for public policy (Counsell, 1998), and the sustainable city has now become a leading paradigm of urban development throughout the world (Whitehead, 2003).

However, while sustainable urban development has been widely accepted, both as a theoretical concept and policy goal, its translation into policy implementation remains problematic. An absence of a working definition of sustainable development has led to the use of the concept of sustainability without any agreement as to its meaning. Furthermore, there has been relatively little analysis of the sustainable city as an object of political contestation and struggle (Whitehead, 2003). Practice from the UK and elsewhere suggests that, at the level of generality the imperative of sustainable development seems unquestionable, but its elaboration within policy permits potentially conflicting and therefore selective interpretations of what constitutes sustainable planning practice (Owen, 1996). Similarly, Drummond and Marsden (1999, p. 10) contend that:

> ... [T]he whole point of sustainable development, the only point which differentiates the concept from narrower ideas such as environmentalism, is that the concept is more than a sum of its parts. It is not just a multi-dimensional concept, it is *fundamentally integrative*. The key problem, however, is that it is profoundly difficult to grasp this multi-dimensionality. Most current approaches focus on and privilege a particular

dimension, be it economic, environmental or social, and what results is often something less than sustainable development.

Within this context, a central urban policy objective is now to reconcile economic competitiveness with social cohesion and environmental protection. In relation to the economy, cities in the Republic of Ireland have been transformed by the impressive economic growth of the 1990s and the so-called 'Celtic Tiger', and in Northern Ireland by the economic 'spin-offs' of the peace process and increased political stability. Between 1993 and 2001, the annual real growth rate of the economy in the Republic of Ireland has been more than double the average recorded over the previous three decades – 8 per cent compared with 3.5 per cent – and there is no doubt that throughout the 1990s, the Republic of Ireland significantly outperformed all other European Union (EU) countries (Clinch et al., 2002). This growth performance during the 1990s and into the start of the present decade has led to rapid convergence of output per capita with the EU average, which has been driven by exceptionally strong growth in employment (ESRI, 2004). The rate of unemployment has as a consequence dropped to historically low levels – from a peak of 17 per cent in the 1980s to under 4 per cent in 2001. This economic growth has been the product of a number of interrelated factors, including: EU membership (and significant EU funding); access to the Single European Market (SEM); sustained US economic growth during the 1990s; the creation of a flexible environment for foreign direct investment (e.g. low corporate tax rates); a high proportion of the population of working age; increased participation in the labour market, particularly among females; sustained investment in education and training; and a series of national social partnership agreements and industrial peace. In Northern Ireland, the impact of the political negotiations and peace process has also brought economic benefits. Although its performance has been much less spectacular than in the Republic of Ireland, Northern Ireland in the 1990s still grew faster than any other region in the United Kingdom (*The Economist*, 2004). Growth has slowed in recent years, but unemployment is down to just over 5 per cent, which is lower than, for example, London, the west Midlands, the North East and Scotland. In terms of demographic profile, Northern Ireland has a regional population growth rate which is twice the current UK rate and exceeds that of the Republic of Ireland, making Northern Ireland one of the fastest growing regions in Europe (DRD, 2001).

However, economic growth has not been evenly distributed throughout the island, with the Greater Dublin Area growing significantly in relative and absolute terms. As an indicator of its economic position within the State, the Greater Dublin Area (GDA) accounted for 46.8 per cent of Gross Value Added[3] for the Republic of Ireland (CSO, 2004). Within the national settlement structure, again Dublin dominates with the GDA having well over twice the population of the next nineteen urban centres outside Dublin (Bannon, 1999). The population of the GDA increased by 60.1 per cent between 1961 and 1998, and as recorded in a background paper for the formulation of the recent National Spatial Strategy:

> Dublin has prospered as a major focus of foreign direct investment into Ireland, as well as an expanding centre of trade and tourism. The city has become a dominant national gateway. At the same time, Dublin functions strongly as the control centre of the economy and of virtually all facets of Irish economic and social life … In many respects Dublin is the centre of Irish economic life (Goodbody Economic Consultants et al., 2000, p. 3).

The success of the Irish economy during the 1990s has been well-documented (see for example, Breathnach, 1998; Clinch et al., 2002; Walsh, 2000) and indeed provides an inviting model for new EU accession countries in Eastern Europe, keen to emulate Ireland's success in achieving rapid economic convergence with the EU average. However, much less international attention has been given to the impact of economic growth on urban patterns of development and social cohesion and urban governance. In this context, there are also valuable insights that can be identified, and lessons that can be learnt, from Ireland's recent experience of rapid economic growth. This book, therefore, provides a timely opportunity to reflect on recent changes in Irish urban areas, particularly relating to both the environmental consequences of urban growth (e.g. urban sprawl) and also to evaluate community cohesion and the increasing role of civil society in managing urban change. The latter is especially difficult in the Northern Ireland context, where despite the peace process, urban communities remain deeply segregated.

The Urban Environment

There is now a growing realisation that much of the sustainability debate has an urban accent. The world's cities are the major consumers of national resources and the major producers of pollution and waste. Any credible strategy to address these problems has to respond to the urban pressures on the environment (McEldowney et al., 2003). Therefore, as Breheny (1992) suggests, if cities can be designed and managed in such a way that resource use and pollution are reduced, then a major contribution to the solution of the global problem can be achieved. A consequence of the sustainable development agenda has been the changing conceptualisation of the urban environment within policy prescription. For example, Davoudi et al. (1996) highlight a number of key trends: firstly, early traditions of urban management were characterised by an 'aesthetic utilitarian' approach which viewed the environment as 'functional resources' to be conserved, and as amenities to be enhanced for human enjoyment and exploitation. Secondly, Davoudi et al. (1996) identify a 'moral and aesthetic notion of the environment as backcloth and setting' to the city. This approach promoted the sharp contrast between town and country and the idea of accommodating growth in a framework of open spaces and a pleasing environmental setting. Thirdly, the 1980s were dominated by the 'marketised utilitarian' approach, whereby the environment is treated as a commodity, a stock of assets that can be priced, traded and packaged for potential investors. However, increasingly urban management is characterised by resource protection approaches which emphasises the concept of capacity in two senses – environmental capacity, the ability of the environment to accommodate development without damage; and urban capacity, generally applied to accommodating development within existing urban areas (Counsell, 1999).

In recent years, the concept of sustainable development has become central in the formulation of spatial and urban management strategies throughout Europe. Since the United Nations Conference on Human Settlement in Rio de Janeiro in 1992, the idea of sustainability has begun to permeate much of central government legislation and policy in European countries. At a European level, the sustainable urban development agenda has gained a high political currency since the early 1990s. In 1990, the Commission of the European Communities (CEC) published the 'Green paper of the Urban Environment', providing a comprehensive review of the challenges facing the urban environment. This was developed further in the 1997 CEC Communication 'Towards an Urban Agenda in the European Union (EU)', which stressed the need for an urban perspective in EU Policies. This was followed by the 1998 Communication 'Sustainable Urban Development in the European Union: A Framework for Action', which outlines four interdependent policy aims that provide a holistic perspective on sustainable development: (1) strengthening economic prosperity and employment in towns and cities; (2) promoting equality, social inclusion and regeneration in urban areas; (3) protecting and improving the urban environment towards local and global sustainability; and (4) contributing towards good governance and local empowerment. Furthermore, this Communication document outlines a series of policy themes to promote sustainable urban development as follows (see also CEC, 2004):

- improve ambient air quality in urban areas, the reliability and quality of drinking water supplies, the protection and management of surface and ground water;
- reduce at source the quantity of water requiring final disposal and reduce environmental noise;
- protect and improve the built environment and cultural heritage, and promote biodiversity and green space with urban areas;
- promote resource efficient settlement patterns that minimise land-take and urban sprawl;
- minimise the environmental impacts of transport through aiming at a less transport-intensive path of economic development and by encouraging the use of more environmentally sustainable transport modes;
- improve environmental performance of enterprises by promoting good environmental management in all sectors;
- achieve measurable and significant reductions of greenhouse gas emissions in urban areas, especially through the rational use of energy, the increased use of renewable energy sources and the reduction of waste;

- minimise and manage environmental risks in urban areas;
- promote more holistic, integrated and environmentally sustainable approaches to the management of urban areas, within functional urban areas and foster eco-systems-based approaches that recognise the mutual dependence between town and country, thus improving linkage between urban centres and their rural surroundings.

Clearly, as Whithead (2003, p. 1186) argues, from the official orthodoxies of sustainable urban development, "it appears that the sustainable city is as much a political vision or social ideal – incorporating its own moral geography and forms of ecological praxis – as it is a tangible object, or location on a map". In the case of Ireland, the rhetoric of sustainability has been embraced in a raft of spatial and urban policy initiatives; however, major questions arise in terms of translating policy objectives into implementation. These issues are addressed in Part One of this book.

Restructuring Urban Governance

Urban revitalisation in many advanced capitalist societies in the 1980s was dominated by a dependence on physical renewal and flagship projects. The central themes of urban regeneration during this period are summarised by Gaffikin and Morrissey (1999b, p. 116) as:

> … the relative emphasis on *downtown* compared to neighbourhood; the fashion for *waterfront development*; the regenerative role of *services*; the increased influence of the private sector, sometimes via new forms of *partnerships*; and the greater use of *place marketing* (original emphasis).

However, since the 1990s the agenda has moved forward and the new urban policy approach is commonly built around the three themes of competitiveness, cohesion and governance (Buck et al., 2002), including an enhanced appreciation for the role of civil society (Lovan et al., 2004). Within this context, recent years have witnessed a radical restructuring of urban and local governance in Ireland, and indeed most advanced capitalist societies. This has included a growing shift from government to governance, the emergence of partnerships as a mechanism for local development, and the imperative to include multiple stakeholders in collaborative models of policy formulation.

In the Republic of Ireland, this re-structuring of governance relationships at a local level emanates from a number of sources (Broaderick, 2002): a new national context for governance that proceeds on the basis of social partnership; European and global influences that emphasise the importance of multi-stakeholder involvement; and increased community development activities in Irish society. This national framework has enabled a radical restructuring of governance at a local level, which has witnessed a changing landscape with a bearing on traditional local representative democracy. The traditional local government system in Ireland, compared to other European states, is relatively weak with a more limited range of functions and powers (Meldon et al., 2004). The system comprises 114 administrative units spread over four categories: town councils, county councils, borough councils and city councils. The principal services provided by the local authorities include: housing and building; road transportation and safety; water supply and sewerage; development incentives and controls, including planning; environmental protection; recreation and amenities; and agriculture, education, health and welfare (though a strictly limited role) (for a detailed account of the role of local government in Ireland, see Callanan, 2003). The weaknesses of local government have been well documented over the years, not least by the Government appointed Barrington Committee in 1991, which outlined:

- A limited range of functions, principally to do with physical development;
- A heavy financial dependency on central government in the absence of a local tax base;
- A marginal role for elected representatives, mainly that of haranguing officials to provide favours for clients (also used as a stepping stone to national politics);
- Fewer units of local government and elected representatives than similar smaller EU member states;
- An inward looking and bureaucratic culture, with little orientation towards the general public or community groups.

However, the terrain of local government has been radically altered through the emergence of local partnerships in the late 1980s and early 1990s. Partnerships have become widely established as a delivery mechanism for local development policy, encompassing membership of public, private and community sectors interests. Partnerships as a delivery mechanism for public policy have become widely established in advanced capitalist societies (Edwards et al., 2001) and is a recognition of complex economic, social and political changes, which have transformed the manner in which policy is made and delivered (Greer, 2001; Scott, 2004). This includes an increasing acknowledgement of the multi-faceted nature of public policy and administration, the inter-connectedness of regeneration decisions taken at the local, regional and international levels (Hart and Murray, 2000), and the increasing fragmentation of policy delivery.

By the end of the 1990s, Walsh (1998) estimates that there were approximately 100 officially recognised local partnerships in the Republic of Ireland, operating in response to a variety of national and EU programmes. These included: County Enterprise Boards, the Area-Based Response to Long Term Unemployment, and European programmes such as LEADER, URBAN and Poverty 3. However, during the mid 1990s a number of tensions emerged concerning the relationship of local development partnerships and local government. For example, concerns were expressed related to public accountability and local democracy, when some commentators raised the possibility of the corrupt use of funds by partnerships, and the role of elected representatives vis-à-vis community nominees became increasingly questioned (see Walsh, 1998; Shortall and Shucksmith, 1998). A further issue was the lack of coordination that was apparent in the activities of the various partnerships operating in the same areas and with the relevant local authority.

The issue of citizen participation in local government, the relationship between local government and local development partnerships, and the lack of coordination of local development activities, has led Government to introduce a number of measures to address these concerns. The commitment for reform of local government in the Department of Environment and Local Government in the Republic of Ireland, produced the White Paper, 'Better Local Government, A Programme for Change' (1996). A core aim of this report was to enhance local democracy ensuring that (p. 10):

- Local communities and their representatives have a real say in the delivery of the full range of public services locally;
- New forms of participation by local communities in the decision-making processes of local councils are facilitated;
- The role of councillors in running local councils is strengthened (including through a widening of the remit of local government, p. 16).

Better Local Government outlined a modernisation programme for local government based on three key themes. Firstly, the White Paper proposed to enhance local democracy through introducing new governance structures. These reform measures included establishing Strategic Policy Committees (SPCs) and Corporate Policy Groups (CPGs) within the local authority system to assist in the formulation, development and review of policy relating to strategic statutory functions. In addition, Area Committees were recommended to strengthen the role of councillors as community representatives and to enable the decentralisation of decision-making to electoral area level. Secondly, Better Local Government proposed a wider role for local government, primarily to be achieved by strengthening the relationship between local government and local development structures. Thirdly, the White Paper recommended a suite of organisational management initiatives designed to improve the quality of services. Measures introduced included one stop shop centres, customer consultation, performance indicators, and human resource management initiatives.

The Better Local Government Programme was further enhanced with the initiation of an Interdepartmental Task Force on the Integration of Local Government and Local Development Systems in 1998, which recommended the establishment of County or City Development Boards (CDBs). These boards are comprised of a partnership of local government, local development bodies (Area-based Partnership Companies, ADM-supported community groups,[4] County/City Enterprise Boards, and LEADER Groups), the social partners including the community and voluntary sector, and representatives of relevant State Agencies at local level. The primary function of the CDBs is to develop and implement a strategy for economic, social and cultural development, and to operate autonomously but under the local government umbrella (Interdepartmental Task Force on the Integration of Local Government and Local Development Systems, 2000).

In Northern Ireland, for a political system where consensus, pluralism and local accountability could only thinly be recognised, the concept of sustainable partnership governance is difficult to contemplate (Murtagh, 2001). However, since the early 1990s, there has been a proliferation of partnership activity in response to new modes of governance, European funding programmes and the wider peace process, as Murtagh highlights in Chapter 10. Area-based partnership formation has been endorsed by a range of Northern Ireland Government Departments and agencies, and towards the end of the 1990s included: 26 Economic Development Partnerships (in each of the District Council areas); 26 District Partnerships (funded by the Peace and Reconciliation Programme and focused on social inclusion); 9 Area-Based Strategy Partnerships for rural development (funded by the Department of Agriculture); 15 LEADER local action groups; 5 urban regeneration partnerships in Belfast (funded by the Department of Social Development); and the emergence of community safety partnerships. In a review of local development in Northern Ireland, Hart and Murray (2000) suggest a number of positive features relating to the emergence of partnerships such as: an enabling institutional framework; new geographies of collaboration; a vibrant culture of participative democracy; the potential for innovative actions; and a contribution to Targeting Social Need. However, Hart and Murray also identify a number of serious deficiencies in partnership-based activity. These limitations relate to: institutional fragmentation; a crowded arena for the delivery of local development; start-stop capacity building; and an over-dependency on short-term public funding. These weaknesses notwithstanding, partnership formation for the delivery of public policy has continued in Northern Ireland into the present decade, suggesting the importance and need for the continuous reflection on current practice.

In parallel to these changes in policy formulation and delivery, new relationships are being forged between local people and the places within which they live and work (Murray and Greer, 1997). In the context of globalisation processes, 'community' often represents a more meaningful spatial scale for individuals to influence action and to manage environmental change (in a world of trading blocs and Multi-National Corporations). Given the increasing retreat of the state in recent years, Low (1999) contends that 'civil society' is often recommended as a form of mediation to replace the politics of mass parties, corporatist organisations, and interest groups which have characterised politics in Western countries for most of the 20th Century. The growth of civil society implies an active citizenship comprising of community groups or associations, where participation is central to local democracy. Although policy-makers have used terms such as 'community' in the past, Friedmann (1998) argues that the term was typically used in a passive, general sense. However, as a political concept, civil society can act as a counterpole to the state, with communities and citizens who not only assert the right to hold the state accountable, but also the right to claim new rights for themselves. In this context, civil action can be reactive or proactive and can take the form of protest or collaboration. Therefore, from this perspective, an increasing interest in community participation can be viewed to be as much a response by government to contemporary community needs as an interest stimulated by any ideological position (Curry, 2001).

Although partnership structures have become the dominant mechanism in terms of policy formulation and implementation, Broaderick (2002) questions the relationship of partnership and participation, which are often seen as synonymous. Broaderick suggests that community and voluntary sector interests have been asked to participate in policy discussion as if it constituted a homogeneous interest. Partnerships also tend to favour those community interests that are already well organised and able to participate in the partnership process (Scott, 2004). Moreover, as Gaffikin and Morrissey (1998) argue, there is often a trade-off between partnership and participation, whereby those suffering most from exclusion are the very ones excluded from this process. At best, Walsh (1997) suggests that partnership redistributes power among a small elite, while participation shares power among a broader constituency. Much less clear is the ability, capacity and opportunities for the more informal networks of residents' groups and civil society organisations at a neighbourhood level to participate in local governance. These issues are addressed in Part Two of this book.

Spatial Policy in Ireland

Recent years have witnessed unprecedented interest in Europe in the formulation of spatial strategies for territorial development (Faludi, 2001; Healey et al., 1997; McEldowney and Sterrett, 2001; Shaw et al., 2000). As recorded by Albrechts et al. (2003), the motivations for these new efforts are varied, but the objectives have typically been to articulate a more coherent spatial logic for land-use management, resource protection, and investments in regeneration and infrastructure. Typically, therefore, spatial planning frameworks embrace a wider agenda than regulatory approaches to land-use management in an attempt to secure integrated policy delivery and more effective linkages between strategic and local planning. The driving forces behind this high profile for spatial strategies have been outlined by Albrechts et al. (2003, p. 115) and include:

- The 'competitiveness' agenda, positioning regions in a European and global economic space;

- Socio-cultural movements and lifestyle changes that focus voter and lobby group attention on environmentally sustainable resource management and the quality of life/environment in places;
- The reassertion of regional and local identity and image formation in the face of globalisation and the European integration project;
- The search for new modes of multi-level governance and a government reorganisation agenda involving decentralisation and the formation of alliances.

This current enthusiasm for strategic spatial planning undoubtedly owes much to the completion of the European Spatial Development Perspective (ESDP) in 1999 and its subsequent political endorsement (Murray, 2003). Through the EU spatial planning process a new discourse of European spatial development is taking shape, with the definition of a new policy discourse, new knowledge forms and new policy options (Richardson, 2000). ESDP promoted concepts – such as polycentric urban development, balanced spatial development, a new urban-rural relationship and transnational planning – are being increasingly translated and applied into individual member state's national and regional policies and strategies. As Faludi (2001, p. 633) argues:

> As a strategic document, the ESDP wants to be 'applied' rather than 'implemented'. Rather than giving shape to spatial development, application is the shaping of the minds of the actors in spatial development.

Following these trends in Europe, significant spatial planning initiatives have been undertaken in Ireland, including the publication of the Regional Development Strategy (RDS) for Northern Ireland (2000-2025) by the Department of Regional Development (DRD) in 2001, and in the Republic of Ireland, the publication of the National Spatial Strategy (2002-2020) (NSS) in 2002 by the Department of Environment and Local Government (DOELG). The two strategies are clearly influenced by the European Spatial Development Perspective, both conceptually and in adopting the EU spatial planning discourse, although interestingly each was distinguished by contrasting planning processes. One of the almost unique features in developing the Northern Ireland strategy was an extensive consultation exercise or participatory planning approach (McEldowney and Sterrett, 2001), involving over 500 community or interest groups in the plan formulation. Although this process has been criticised (see for example, Neill and Gordon, 2001), Murray and Greer (2003) and Healey (2004) contend that this approach has been highly inclusive, standing in contrast to the previous expert dependent and technocratic prescriptions of past regional planning, providing legitimisation for the current framework. In contrast, the formulation of the National Spatial Strategy in the Republic of Ireland was marked by a politically-driven process. Preparation work on the NSS commenced in January 2000 and, while its publication was anticipated in late 2001, a General Election during mid 2002 delayed its release until the end of the year. Planning is very much a political activity and thus the sensitivities attached to the possible designation (and non-designation) of growth centres would undoubtedly have placed the spatial strategy at the centre of political controversy in the run-up to voting day (Murray, 2003).

The broad aim of the Northern Ireland spatial strategy is to guide future development in order to "promote a balanced and equitable pattern of sustainable development across the Region" (p. 41) and a framework of interconnected hubs, corridors and gateways is adopted (see Figure 1.2). The strategy aims to achieve a balance of growth to maintain a strong economic heart in the wider Belfast travel-to-work hinterland while encouraging decentralised development at identified growth poles across Northern Ireland. This will be focused on the North-West and the main urban centres throughout rural Northern Ireland. The objective of the hub, corridor and gateway approach is to provide a strategic focus to future development and achieve balanced growth by developing (DRD, 2001, p. 43):

- The key and link transport corridors as the skeletal framework for future development;
- A compact and dynamic core centred on Belfast, the major regional gateway;
- A strong North-West regional centre based on Derry/Londonderry;
- A vibrant rural Northern Ireland with balanced development spread across a polycentric network of hubs/clusters based on the main urban centres.

Similarly, the Republic of Ireland's National Spatial Strategy is intended to provide for the first time, an explicit national framework for dealing with spatial issues. The NSS sets out a twenty-year planning framework designed to achieve a better balance of social, economic, physical development and population growth on an inter-regional basis and to set a national context for spatial planning including the preparation of regional planning guidelines and county and city land-use development plans. The spatial strategy is comprised of three key elements (see Figure 1.3). Firstly, the NSS aims to promote a more efficient Greater Dublin Area, which continues to build on its competitiveness and national role. However, it is recognised that it is not desirable for the city to continue to spread physically into the surrounding counties. Therefore, the NSS proposes the physical consolidation of Dublin supported by effective land-use policies for the urban area, such as increased brownfield development, and a more effective public transport system.

Secondly, the Strategy designates strong 'gateways' in other regions. Balanced national growth and development is to be secured with the support of a small number of nationally significant urban centres which have the location, scale and critical mass to sustain strong levels of job growth in the regions. The National Development Plan 2000-06 had previously designated Cork, Limerick/Shannon, Galway and Waterford as gateways, and the NSS identified four new national level gateways: Dundalk, Sligo, and two 'linked' gateways of Letterkenny (linked to Derry in Northern Ireland) and Athlone/Tullamore/Mullingar (see Figure 1.3).

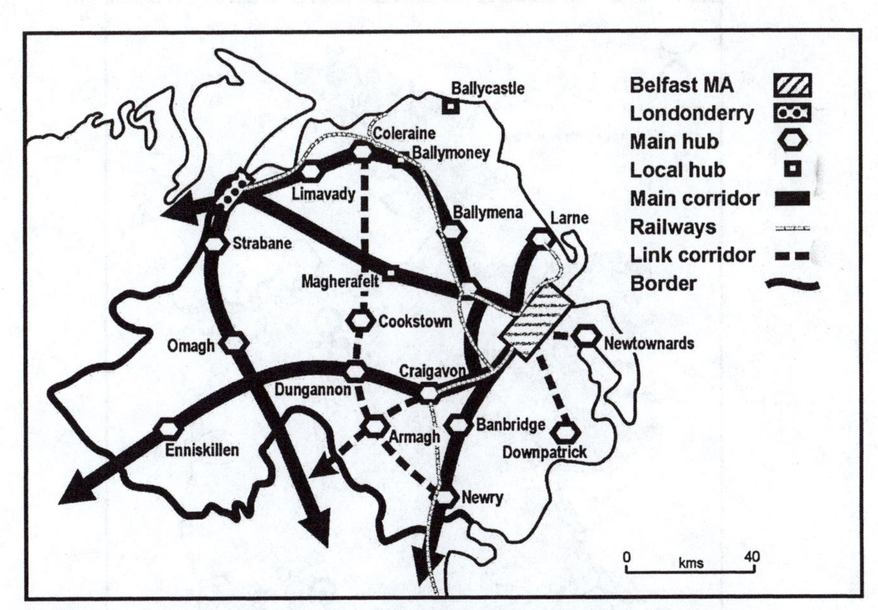

Figure 1.2 Regional development strategy for Northern Ireland

Undoubtedly the designation of gateways was underpinned by political pragmatism. The gateways originally designated in the National Development Plan, with the exception of Galway, are located in the south and east of the State, which are the most prosperous regions in the Republic of Ireland. The designation of the four new gateways in the NSS allows for a more inclusive process, involving the border, midlands and western regions. However, questions can be raised concerning whether these latter four gateways have the critical mass to secure balanced regional development; for example, Dundalk and Sligo have populations of approximately 32,000 and 19,000 respectively, compared to Cork and its hinterland with 350,000 and Limerick and its hinterland with 236,000. This would imply that considerable development is required in this second tier of gateways before they contribute to balanced regional development. Furthermore, the number of gateways designated (eight in total) may prove too many to effectively develop clusters of economic growth and agglomerations, which have access to large labour markets and sub-supply sectors necessary to counterbalance the dominance of the Greater Dublin Area. This has been further weakened by the Government's decision to decentralise the Civil Service. Rather than selecting centres for relocation on the basis of gateways and hubs designated in the NSS, a policy of dispersal has emerged, which includes small rural towns, many would argue driven by political expediency before an election year. This programme will involve over 10,000 civil servants and eight Government Departments relocating from Dublin to 53 centres in 25 counties.

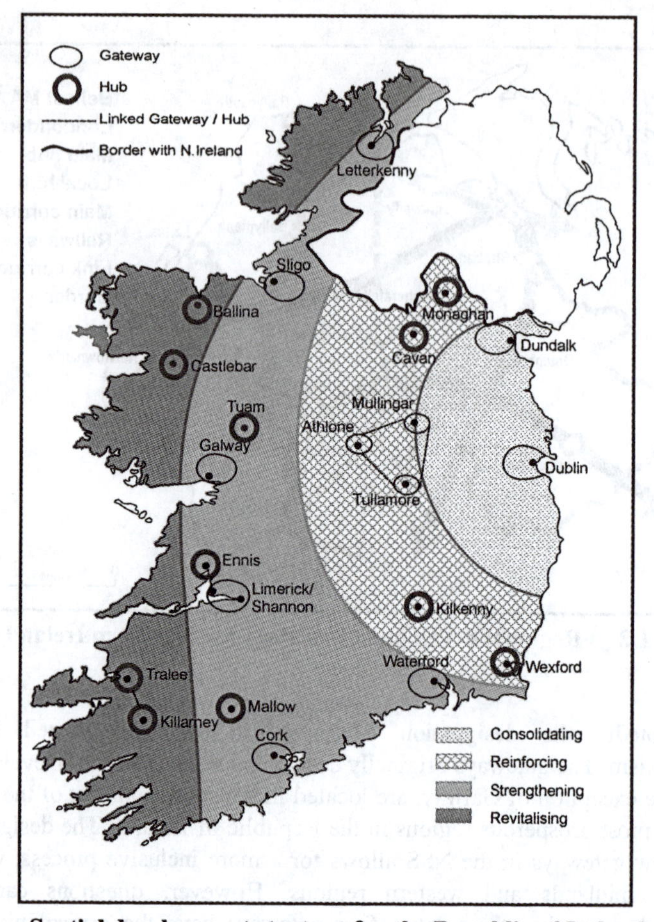

Figure 1.3 Spatial development strategy for the Republic of Ireland

Thirdly, the Strategy also identifies nine medium sized 'hubs', which are to support and be supported by the gateways and will link out to wider rural areas. The hubs identified include Cavan, Ennis, Kilkenny, Mallow, Monaghan, Tuam and Wexford and two linked hubs comprised of Ballina/Castlebar and Tralee/Killarney. These three elements are complemented in the Strategy by mention of the need to support the county and other town structure and to promote vibrant and diversified rural areas. The settlement hierarchy is further developed with its relationship to the proposed national transport framework based on radial corridors, linking corridors and international access points. As recorded by Murray (2003), the NSS is very much skeletal in design and thus in terms of implementation, it is acknowledged that further work as being necessary. In this regard, provisions were made in the recent Planning and Development Act 2000 (Government of Ireland, 2000) for the State's eight Regional Authorities to prepare statutory Regional Planning Guidelines (RPGs) to give full effect to the principles outlined in the NSS. The Regional Authorities were established in 1994, however, their responsibilities have been limited and public profile has been poor. Again, similar to the formulation of a national spatial framework, RPGs are a new instrument in the Irish planning system, and perhaps provide an opportunity to assist in invigorating regional governance in Ireland. In this regard, strategic planning guidelines covering the Dublin and Eastern Regional Authorities, published in advance of the NSS have subsequently been revised in light of the new national framework, summarised in Figure 1.4.

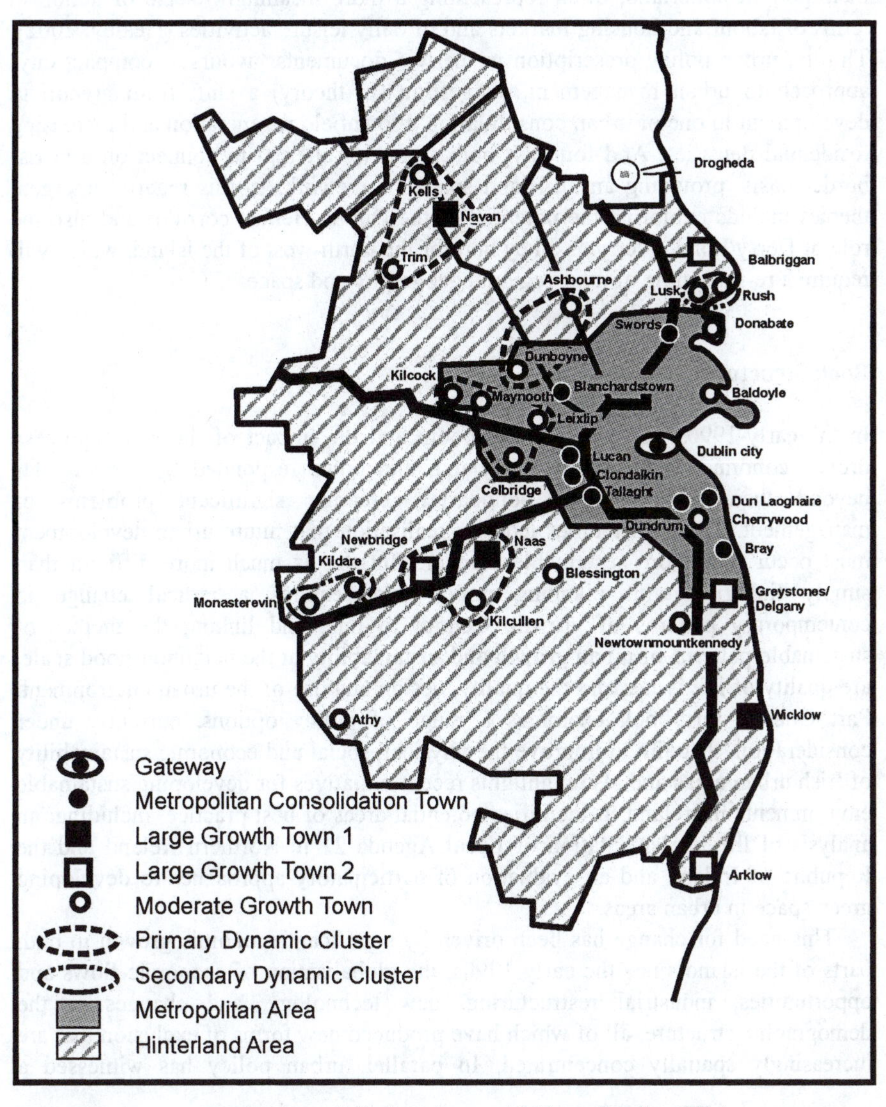

Figure 1.4 Revised spatial planning guidelines for the GDA

While differences exist between the respective strategies for Northern Ireland and the Republic of Ireland, a number of common themes can be identified suggesting a degree of policy convergence between the two parts of the island. Firstly, both strategies signal that planners and plans in Ireland are increasingly moving beyond 'land-use' planning to embrace a wider agenda, adopting in effect the European notion of spatial planning instead of the traditional UK and Irish regulatory approach. This new thinking seeks to achieve an integration of social, economic and environmental dimensions in place-making processes. Secondly, both strategies attempt to broker connections between the core city and its wider metropolitan hinterland, often representing a more meaningful scale of action in terms of labour and housing markets and of daily leisure activities (Healey, 2002). Thirdly, urban policy prescription in the two documents favours a 'compact city' approach to urban management, suggesting (in theory) a shift from greenfield development to one of urban consolidation, brownfield regeneration and increasing residential densities. And fourthly, both strategies attempt to connect on a cross-border basis, providing an all-island spatial framework. In this regard, emergent themes include exploring the potential of the Dublin-Belfast corridor and also the role of Derry/Londonderry as a key hub for the north-west of the island, which will require a re-imagining of inter-regional geography and space.

Book Structure

In the early-1990s, the Celtic Tiger boom and the impact of the peace process drove economic development in Ireland, and cities responded by growing far beyond their administrative boundaries, creating significant problems of management. These difficulties led to a realisation that future urban development must occur in a more sustainable manner. But this is much more difficult than simply tinkering with planning policy and requires a radical change in contemporary patterns of growth. Central to this, and linking the themes of sustainable patterns of urban growth and sustainability at the neighbourhood scale, are quality of life issues and the quality, design and use of the urban environment. Part One of this book examines a range of policy options, currently under consideration, that aim to improve the physical, social and economic sustainability of Irish urban areas and also highlights recent initiatives for developing sustainable environments in Ireland. It identifies potential areas of best practice, including: an analysis of the implementation of Local Agenda 21 in Northern Ireland and the Republic of Ireland and an evaluation of participatory approaches to developing green space in urban areas.

This need for change has been driven by significant economic growth in both parts of the island since the early 1990s, the globalisation of economic flows and opportunities, industrial restructuring, new technology, and changes in the demographic structure, all of which have produced new forms of exclusion that are increasingly spatially concentrated. In parallel, urban policy has witnessed a turn to social objectives with a restructuring of governance relationships and an increased emphasis on communities and collaborative policy formulation. Part Two of the book begins with a comparative analysis of the implementation of Local Agenda 21 in Northern Ireland and the Republic of Ireland, while the remainder of the section presents newly researched case studies to illustrate this shift in urban policy. The themes of partnership governance, active citizenship and the role of residents' associations, planning for neighbourhoods divided by ethno-religious conflict, and affordable housing are examined.

Notes

[1] The Greater Dublin Area is composed of the four local authority areas comprising the Dublin Regional Authority Area (Dublin City, Dublin South, Dun Laoghaire Rathdown and Fingal) as well as those counties adjacent to Dublin that constitute the Mid-East Region (Kildare, Meath and Wicklow).
[2] The Belfast Metropolitan Area comprises an extensive built up area stretching along the shores of Belfast Lough and up the Lagan Valley, taking in the city of Belfast and the adjacent urban parts of the district council areas of Carrickfergus, Castlereagh, Lisburn, Newtownabbey and North Down.
[3] Gross Value Added is a measure of the value of goods and services produced in a region. It includes the profits derived from the production of goods and services and the profits of multinational companies.
[4] ADM or Area Development Management is an EU and national government funded intermediary agency set up in 1992 to promote social inclusion, reconciliation and equality and to counter disadvantage through local social and community development.

References

Albrechts, Healey, P. and Kunzmann, R. (2003) 'Strategic Spatial Planning and Regional Governance in Europe', *Journal of the American Planning Association*, Vol. 69, pp. 113-129.

Bannon, M. (1999) 'The Need for Vision and Strategy in Irish Planning', *Pleanáil*, No. 14, pp. 1-6.

Barrington Committee (1991) *Local Government Reorganisation and Reform – Report of the Advisory Expert Committee*, Dublin, Government Publications.

Boal, F. (1995) *Shaping a city: Belfast in the late twentieth century*, Belfast: Institute of Irish Studies, Queens University Belfast.

Breathnach, P. (1998) 'Exploring the 'Celtic Tiger' Phenomenon: causes and consequences of Ireland's economic miracle', *European Urban and Regional Studies*, Vol. 5, pp. 305-316.

Breheny, M. (1992) 'The Contradictions of the Compact City: A Review', in Breheny, M. (ed.) *Sustainable Development and Urban Form,* London: Pion Limited.

Briassoulis, H. (1999) 'Who Plans Whose Sustainability? Alternative Roles for Planners', *Journal of Environmental Planning and Management*, Vol. 42, pp. 889-902.

Broaderick, S. (2002) 'Community development in Ireland – a policy review', *Community Development Journal*, Vol. 37, pp. 101-110.

Callanan, M. (2003) 'The role of local government', in Callanan, M. and Keogan, J. (eds.) *Local Government in Ireland: Inside Out*, Dublin: IPA.

Carroll, A. (2004) *The Ecological Footprint of Dubliners*, Unpublished MA Thesis, Department of Geography, University College Dublin

Clinch, P., Convery, F. and Walsh, B. (2002) *After the Celtic Tiger, Challenges Ahead*, Dublin: O'Brien Press.

Commission of the European Communities (1990) *Green Paper on the Urban Environment*, Brussels: CEC.

Commission of the European Communities (1997) *Towards an Urban Agenda for the European Union*, COM (1997) 197 final, Brussels: CEC.

Commission of the European Communities (1998) *Sustainable Urban Development in the European Union: A Framework for Action*, COM (1998) 605 final, Brussels: CEC.

Commission of the European Communities (2004) *Towards a Thematic Strategy on the Urban Environment*, COM (2004) 60 final report, Brussels: CEC.

Committee for Spatial Planning (1999) *European Spatial Development Perspective: Towards a balanced and sustainable development of the territory of the EU*, Luxembourg: CEC.

Cooper, J., Ryley, T. and Smyth, A. (2001) 'Contemporary lifestyles and the implications for sustainable development policy: lessons from the UK's most car dependant city, Belfast', *Cities*, Vol. 18, pp. 103-113.

Counsell, D. (1998) 'Sustainable development and structure plans in England and Wales: a review of current practice', *Journal of Environmental Management and Planning*, Vol. 41(2), pp. 177-194.

Counsell, D. (1999) 'Making sustainable development operational', *Town and Country Planning*, Vol. 68(4), pp. 131-133.

CSO (Central Statistics Office) (2004) *County Incomes and Regional GDP 2001*, Dublin: CSO.

Curry, N. (2001) 'Community Participation and Rural Policy: Representativeness in the Development of Millennium Greens', *Journal of Environmental Planning and Management*, Vol. 44, pp. 561-576.

Davoudi, S., Hull, A. and Healey, P. (1996) 'Environmental concerns and economic imperatives in strategic plan making', *Town Planning Review*, Vol. 67, pp. 421-436.

DOE (Department of Environment (2001) *Belfast Metropolitan Area Plan 2015 Issues Paper,* Belfast: DOE.

DOELG (Department of Environment and Local Government) (1996) *Better Local Government, A Programme for Change*, Dublin: DOE.

DOELG (Department of Environment and Local Government) (2002) *The National Spatial Strategy 2002-2020, People, Places and Potential*, Dublin: Stationary Office.

DRD (Department of Regional Development) (2001) *Shaping Our Future – Regional Development Strategy for Northern Ireland 2025*, Belfast: DRD.

Drummond, I. and Marsden, T. (1999) *The Condition of Sustainability*, Routledge: London.

Edwards, B., Goodwin, M., Pemberton, S. and Woods, M. (2001) 'Partnerships, power, and scale in rural governance', *Environment and Planning C: Government and Policy*, Vol. 19, pp. 289-310.

Ellis, G. and McKay, S. (2000) 'City Management profile, Belfast', *Cities*, Vol. 17, pp. 47-54.

ESRI (Economic and Social Research Institute) (2004) *Irish Economy Overview*, accessed at www.esri.ie , November 16th 2004.

Faludi, A. (2001) 'The Application of the European Spatial Development Perspective: Evidence from the North-West Metropolitan Area', *European Planning Studies*, Vol. 9, pp. 663-675.

Friedmann, J. (1998) 'Planning Theory Revisited', *European Planning Studies*, Vol. 6, pp. 245-253.

Gaffikin, F. and Morrissey, M. (1999a) 'The urban economy and social exclusion: The case of Belfast', in Gaffikin, F. & Morrissey, M. (eds) *City Visions, Imaging Place, Enfranchising People,* London: Pluto Press, pp. 34-57.

Gaffikin, F. and Morrissey, M. (1999b) 'Urban regeneration: The new policy agenda', in Gaffikin, F. & Morrissey, M. (eds) *City Visions, Imaging Place, Enfranchising People,* London: Pluto Press, pp. 116-150.

Goodbody Economic Consultants, Department of Regional and Urban Planning (UCD) and The Faculty of the Built Environment (UWE) (2000) *The Role of Dublin in Europe*, A Report Prepared for the Spatial Planning Unit, Department of Environment and Local Government, Dublin.

Greer, J. (2001) 'Whither partnership governance in Northern Ireland?', *Environment and Planning C: Government and Policy*, Vol. 19, pp. 751-770.

Greer, J. & Murray, M. (2003) 'Rethinking Rural Planning and Development in Northern Ireland', in Greer, J. and Murray M. (eds) *Rural Planning and Development in Northern Ireland,* Dublin: Institute of Public Administration.

Hart, M., Murray, M. (2000) *Local Development in Northern Ireland – The Way Forward*, Belfast: Northern Ireland Economic Council.

Healey, P. (2004) 'The Treatment of Space and Place in the New Strategic Spatial Planning in Europe', *International Journal of Urban and Regional Research*, Vol. 28, pp. 45-67.

Healey, P. (2002) 'Urban-Rural Relationships, Spatial Strategies and Territorial Development', *Built Environment*, Vol. 28, pp. 331-339.

Healey, P., Khakee, A., Motte, A. and Needham, B. (eds) (1997) *Making Strategic Spatial Plans: Innovation in Europe*, London: UCL Press.

Healey, P. & Shaw, T. (1993) 'Planners, Plans and Sustainable Development', *Regional Studies*, Vol. 27, pp. 769-776.

Interdepartmental Task Force on the Integration of Local Government and Local Development Systems (2000) *A Shared Vision for County/City Development Boards*, Dublin: Government Publications.

Lindsey, G. (2003) 'Sustainability and Urban Greenways: Indicators in Indianapolis', *Journal of the American Planning Association*, Vol. 69, pp. 165-180.

Lovan, W., Murray, M. and Shaffer, R. (2004) 'Participatory Governance in a Changing World', in Lovan, W., Murray, M. and Shaffer, R. (eds) *Participatory Governance: Planning, Conflict Mediation and Public Decision-Making in Civil Society*, Aldershot: Ashgate.

Low, M. (1999) 'Their masters' voice: communitarianism, civic order, and political representation', *Environment and Planning A*, Vol. 31, pp. 87-111.

McDonald, F. (2002) 'Comment: On the Road to Ruin', In Convery, F. & Feehan, J. (eds) *Achievement and Challenge: Rio +10 and Ireland*, Dublin: Environmental Institute, University College Dublin.

McEldowney, M., Scott, M. and Smyth, A. (2003) 'Integrating land-use planning and transportation: policy formulation in the Belfast Metropolitan Area', *Irish Geography*, Vol. 36, pp. 112-126.

McEldowney, M. and Sterrett, K. (2001) 'Shaping a Regional Vision: the Case of Northern Ireland', *Local Economy*, Vol. 16, pp. 38-49.

McPeake, J. (1998) 'Religion and Residential Search Behaviour in the Belfast Urban Area', *Housing Studies*, Vol. 13, pp. 527-548.

Meldon, J., Kenny, M. and Walsh, J. (2004) 'Local government, local development and citizen participation: lessons from Ireland', in Lovan, W., Murray, M. and Shaffer, R. (eds) *Participatory Governance: Planning, Conflict Mediation and Public Decision-Making in Civil Society*, Aldershot: Ashgate.

Murray, M. (2003) 'Strategic Spatial Planning on the Island of Ireland: Towards a New Territorial Logic?', Paper presented to conference: *Linking Development with the Environment – The EU and Accession Countries Perspective*, Bratislava, Slovakia, February.

Murray, M., Greer, J. (1997) 'Planning and Community-led Rural Development in Northern Ireland', *Planning Practice and Research*, Vol. 12, pp. 393-400.

Murtagh, B. (2001) 'City Visioning and the Turn to Community', *Planning Practice and Research*, Vol.16, No.1, pp. 9-19.

Murtagh, B. (2001) 'Partnerships and Policy in Northern Ireland', *Local Economy*, Vol. 16, pp. 50-62.

Neill, B. & Gordon, M. (2001) 'Shaping our Future? The Regional Strategic Framework for Northern Ireland', *Planning Theory and Practice*, Vol. 2, pp. 31-52.

Owen, S. (1996) 'Sustainability and Rural Settlement Planning', *Planning Practice and Research*, Vol. 11, pp. 37-47.

Richardson, T. (2000) 'Discourses of Rurality in EU Spatial Policy: The European Spatial Development Perspective', *Sociologia Ruralis*, Vol. 40, pp. 53-71.

Scott, M. (2004) 'Building institutional capacity in rural Northern Ireland: the role of partnership governance in the LEADER II Programme', *Journal of Rural Studies*, Vol. 20, pp. 49-59.

Shaw, D, Roberts, P. and Walsh, J. (eds) (2000) *Regional Planning and Development in Europe*. Aldershot: Ashgate.

Shorthall, S. and Shucksmith, M. (1998) 'Integrated Rural Development: Issues Arising from the Scottish Experience', *European Planning Studies*, Vol. 6, pp. 73-88.

The Economist (2004) *The Luck of the Irish: A survey of Ireland*, October 16th 2004.

Walsh, J. (1998) 'Local Development and Local Government in the Republic of Ireland: From Fragmentation to Integration?', *Local Economy*, Vol. 12, pp. 329-341.

Walsh, J. (2000) 'Dynamic Regional Development in the EU Periphery: Ireland in the 1990s', in Shaw, D., Roberts, P. and J. Walsh (eds) *Regional Planning and Development in Europe*, Aldershot: Ashgate.

World Commission on Environment and Development (1987) *Our Common Future*, Geneva: OUP.

PART I
SUSTAINABLE URBAN ENVIRONMENTS

Chapter 2

Re-greening Brownfields: Land Recycling and Urban Sustainability

Niamh Moore

Introduction

Although, as Chapter 1 has illustrated, the rhetoric of sustainability permeates much central government policy in Ireland, little has been achieved in a range of areas to actually put the aspirations into practice. Perhaps the greatest barrier to the embedding of the concept within current urban policy is structural, given that it requires an unprecedented holistic approach to land management and development, an integration of all environmental policy areas including those impacting on human health through soil, air, water or groundwater pollution and the courage to embrace a range of actors and techniques within the urban environment. Former industrial, under-utilised or derelict areas provide all of these challenges and quite possibly because of this have been sidelined and dealt with in a piecemeal fashion, regeneration only being perceived as a by-product of, as opposed to the rationale for, broader rejuvenation. Much of the international literature contends that the major barrier to such 'brownfield' regeneration globally is economic, that it is simply too costly to remediate and recycle this land, conveniently ignoring the fact that brownfields have been created because of economic structural change. This realisation could become the first step towards developing an integrated approach to problem-solving, given that it would demand not just environmental, but also social and community solutions to be considered. The goals of orchestrating the competing demands of economic growth, ecological and social vitality at national level and providing the tools to implement them at local level have become perceived as insurmountable. This chapter reviews why such areas have a critical role to play in encouraging urban sustainability through a range of international case studies, and it examines the policy context and progress towards brownfield regeneration in Ireland.

The European Context

In an early effort to promote sustainable urban development in line with the Rio recommendations, the European Sustainable Cities Project (1996) argued that as well as accepting that land is a finite resource, policy should be supply-driven rather than demand-driven, and land recycling encouraged. Recognising that the problems of urban sustainability are most acute in "areas where residential densities are low and where day-to-day activities (home, work, shopping) are widely separated" (European Union, 2001, p. 2), the Union has favoured the adoption of local responses to urban issues. In the Treaty of Amsterdam (1997) sustainable development - assigning equal importance to economic and environmental issues - was explicitly identified as a key strategy for the European Union. The primary recommendations of the more recent Guiding Principles for Sustainable Spatial Development of the European Continent are to control the expansion of urban areas through activating gap sites and increasing the supply of land within cities by recycling previously-used land, revitalising deprived neighbourhoods, and engaging in environmental regeneration of former industrial sites. Although recommendations have been made in many countries to ensure this by adopting a sequential approach to planning that would favour development on brownfield rather than greenfield sites, the difficulties inherent in doing it in different cultural contexts have been widely acknowledged.

However, recent work by the CLARINET Working Group has begun to establish commonalities across Europe. Grimski & Ferber (2001) have begun by identifying three types of brownfield in a European context: those in traditional industrial areas, in metropolitan areas and in rural areas. They argue that although "the three disciplines of environmental restoration, land-use planning and economic policy are all involved in the process of brownfield redevelopment" (2001, p. 143), it appears that structural or economic policy objectives are the most predominant driving forces in regeneration across the continent. Whether redevelopment succeeds, however, is driven by much broader processes. Four common factors underly the success or failure of regeneration, providing major challenges not just in a European, but in a global context (Figure 2.1).

Ireland in International Perspective

Perhaps one of the reasons why Ireland has taken longer to realise both the problem and potential of brownfields is historical. Unlike many other European countries which had their social and economic fabric entirely re-written by the Industrial Revolution, Ireland was affected to a much lesser extent. A common perception appears to abound that this country has made a direct transition from an agriculturally-based to a post-industrial economy, and therefore brownfields as relics of the industrial period are not very extensive. It is true that in comparison to Germany, which has approximately 128,000 hectares of brownfield land or the Netherlands, which has in the region of 9,000-11,000 hectares, the extent of the problem in Ireland is small but this does not mean that lessons cannot be learnt from the experience elsewhere to chart a future direction for Irish spatial policy.

This is particularly important in the context of an emergent approach to

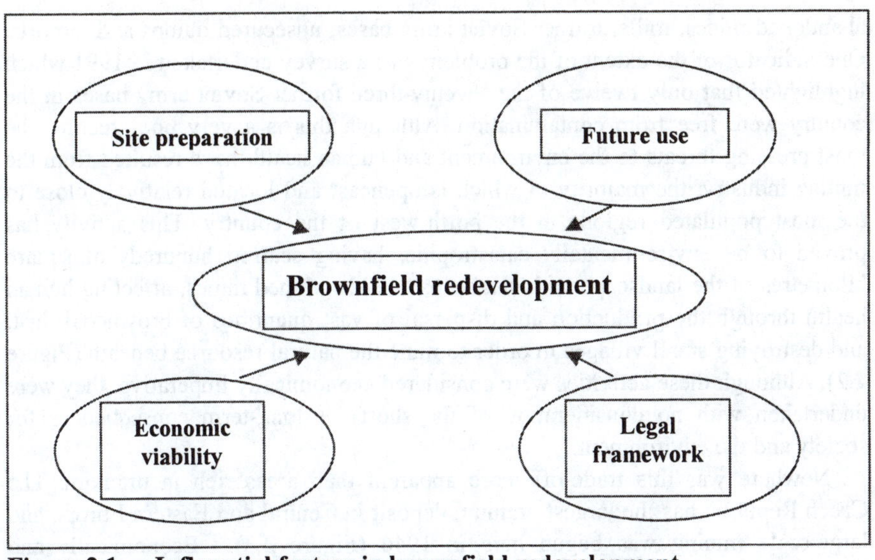

Figure 2.1 Influential factors in brownfield redevelopment
Source: Grimski & Ferber, 2001

development in Ireland, outlined in the Introductory chapter, that is based on an attempt to develop and implement spatial policy in a strategic and holistic manner. Some of the obstacles faced by those attempting to implement this approach are also proving to be barriers to a reconception of the future and role of brownfields in Ireland, discussed in more detail below.

Economic Viability and Legal Framework

Of the four common factors, identified by Grimski and Ferber, that play a major role in influencing the likely outcome of brownfield redevelopment, two are structural issues - economic viability and the legal framework - and thus provide unique challenges in different spatial contexts. This is nowhere more evident that in the former Eastern Europe where the previous socialist economy and the absence of any environmental controls facilitated the emergence of brownfields and proved major inhibiting factors to the regeneration of former industrial areas. During the transition period, environmental regeneration has not been high on the list of priorities, but the recent preparations and actual accession of many of these countries to the European Union with its relatively strict environmental regulations has begun to change this.

One such country has been the former Czechoslovakia (the present-day Czech Republic) where during the Soviet era an exploitative, rather than a symbiotic, relationship developed between society and nature or more simply humans and their environment. The legacy has been catastrophic and today brownfields represent the most serious environmental threats in the Czech Republic from abandoned mines, mills, former Soviet army bases, unsecured dumps and airports. One indicator of the extent of the problem was a survey undertaken in 1991 which highlighted that only twelve of the seventy-three former Soviet army bases in the country were free from contamination. Although this is a very poor record, the most pressing threats to the environment and human health have resulted from the mining industry, the majority of which is opencast and located relatively close to the most populated regions in the north-west of the country. This activity has proved to be environmentally catastrophic, having scarred hundreds of square kilometres of the landscape with slagheaps and abandoned mines, affecting human health through the production and dispersal of vast quantities of browncoal dust, and destroying small villages in order to mine the natural resource beneath (Figure 2.2). Although these activities were considered economically imperative, they were undertaken with no consideration of the short- or long-term consequences for society and the environment.

Nowhere was this trade-off more apparent than areas rich in uranium. The Czech Republic has the largest uranium deposit in Central and Eastern Europe and large-scale mining was begun here in 1946 (Figure 2.3). Economically and militarily this was an important resource and mining reached its maximum

Figure 2.2 Open cast mining in Northern Bohemia

potential in 1960 with 3,000 tonnes mined per annum. In order to source the uranium, over 6,000 wells were constructed over a 600 hectare area and sulphuric acid was pumped into the uranium deposits lying within layers of sandstone. The liquid was then pumped off and the uranium separated from it. Beyond the straightforward environmental consequences of pumping acid into the soil, this deposit also lay beneath the largest groundwater reserve in Northern Bohemia. In effect the sulphuric acid was being pumped into drinking water, creating a major threat to human health.

Since 1992 following the collapse of the Iron Curtain and the 'Velvet Divorce' which resulted in the break up of Czechoslovakia, the Czech government through the Ministry of the Environment has been actively working to clean up over 1500 sites affected by uranium mining activities over the last fifty years. Now under a capitalist regime, the economic rationale no longer exists for leaving these sites

Figure 2.3 Uranium mine at Jachymov, Bohemia

derelict and legal compliance with EU directives as part of accession agreements has resulted in an entirely new attitude and approach to these problem areas.

Site Preparation and Future Use

However, it would be a serious mistake to assume that trade-offs between the economy and the environment do not occur during redevelopment, for it is often during the clean-up of these sites that this kind of bargaining becomes most obvious as has occurred in the Canadian capital, Ottawa. Known locally as The Flats because of the low-lying topography, LeBreton Flats are located one mile west of the city centre, on the south banks of the Ottawa River, adjacent to Victoria Island (see shaded area in Figure 2.4). Although overlooked by Parliament Hill, this area has remained undeveloped since it was cleared in 1965, in marked contrast to the vitality and vibrancy that once characterised the area.

One of the oldest districts in the city, LeBreton Flats became established as an important industrial site in the 1850s, due in large part to its geographical location, close to the Chaudière Falls. It evolved into a timber town in the nineteenth century, a contributory factor in facilitating a major fire which swept across the area in 1900 and left behind coal, ash and a legacy of contamination.

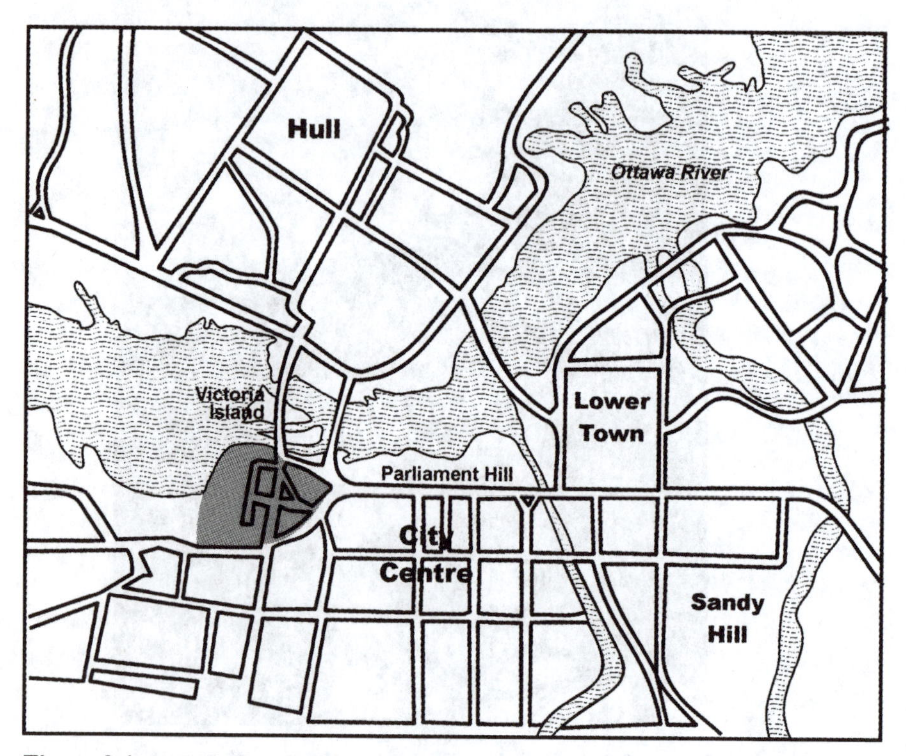

Figure 2.4 LeBreton Flats in relation to the city centre

The site was rebuilt cheaply and quickly became a focus for the primary and fabricated metal industries, scrap yards and some worker's housing were added to the environmental mix. By the 1920s automotive vehicle servicing, storage and wrecking yards. By the 1960s this area had taken on all the characteristics of a zone in terminal decline with poor quality housing cheek-by-jowl with industrial activity. In 1965 a decision was taken to clear the area and for forty years, this sixty-six acres remained derelict in the heart of the city.

One of the key reasons for this was the industrial legacy, which had produced an estimated 246,000 m^3 of contaminated land, the treatment of which was disputed. In the late 1990s, two consultancy reports were commissioned to determine how to handle this problem. The first produced by Aqua Terre Solutions recommended in March 2000 that the government should transport the polluted soil to an offsite landfill, while the second report by IBI Group proposed that the polluted soil be dug up and reburied in the immediate vicinity rather than hauled off to a distant dump. The latter argued that the ideal would be to dump it on the former Nepean Bay landfill, on the bank of the Ottawa River at the western edge of the LeBreton Flats, which could save the government $20 million in cleanup costs and make proposed development more affordable. This again appears to highlight the almost inevitable conflict between economic and environmental concerns, between making the best decision for society / environment or for the economic paymasters. The second proposal overtly highlighted the short-term thinking that permeates much urban development and planning organisations, trying to push the problem somewhere else, in this case into the adjacent river, or into the future. It begs the question that if this is the reputation brownfield regeneration is acquiring and this kind of expediency is driving decision-making, then how can it possibly contribute successfully to the agenda of sustainable development?

Moving Forward: Policy Initiatives

Given the extent of economic restructuring over the last twenty-five years, it makes sense to look to former industrial heartlands for answers to this question. In the United States, a range of policy initiatives has been taken to deal with the restoration of previously-developed land including the introduction of the Superfund. This federal scheme provides the financial and legislative support to engage in regeneration, through financial incentives and subsidies, the provision of liability protection mechanisms and the guarantee of reasonable clean-up standards.

However at present, the best lessons can perhaps be learned from the UK, which has a similar legislative system to Ireland and appears to be successfully introducing a range of policies to deal with brownfield areas. Since 1997, responsibility for environmental, and very specifically brownfield, policy has been vested in the Office of the Deputy Prime Minister (ODPM). In the original renewal programmes in Britain, an emphasis on new industry for previously-developed land was highly apparent but given the changing nature of the world economy this provided limited opportunities for innovation. In recent years, government policy has encouraged re-development for mixed-use activities, with an emphasis on reaching a target of constructing 60 per cent of all new housing on brownfield land by 2008. By 2002, 64 per cent of all new housing developments, including conversions, were being built on brownfields (Office of the Deputy Prime Minister), facilitated through a number of policy initiatives, chief among which was the establishment of a National Land Use Database.

This partnership project between the Office of the Deputy Prime Minister, English Partnerships, the Improvement and Development Agency (representing the interests of local government) and Ordnance Survey, collects data on vacant and derelict sites and other previously developed land and buildings that may be available for redevelopment. The second element is the production of an up-to-date large-scale land use map that may be used in future years as a baseline study map. This land use map has also begun to prove important in making zoning decisions and targeting areas for strategic investment.

Additionally, and to complement this initiative, a partnership of four national organisations in England; Groundwork, English Partnerships, the Forestry Commission and the Environment Agency are developing the Land Restoration Trust. The overall aim is to improve the quality of the environment through the acquisition of derelict land that is not suitable for redevelopment and would otherwise remain an environmental and aesthetic problem. The Trust will work with local organisations to create new 'green amenities' providing benefits both for society and the natural and built environment. This bottom-up approach to brownfield regeneration complements the top-down National Land Use Database and fosters confidence in the potential of former industrial sites.

Other policy measures emanating from and encouraged by the ODPM have been the introduction of fiscal incentives in Budget 2001; reform of urban renewal legislation; guidance for local authorities and developers on unlocking the potential of empty properties and the introduction of pathfinder projects to help local authorities tackle low demand and abandoned homes where the problems are most acute. Yet even with all of these policy tools in place, there is still an argument made that much more could be done:

> Citizens do see the link between the quality of their lives and the quality of civic leadership, but feel totally powerless to affect change. They don't know where to go to get an answer. Britain needs a national framework that enables and promotes Urban Renaissance, and local political structures that can deliver it … Favour brownfield first – provide a level playing field through fiscal and legal strategies to make brownfield more attractive than green. This should allow us to increase the percentage of brownfield development from 60 to 75 per cent in the first instance (Richard Rogers speaking to Urban Summit, 2002).

This continuous raising of the bar as to what is possible and best practice in achieving it, provides lessons to Irish policy-makers and decision-makers who up until now have been slow to embrace this process as a potential revolutionary step for Irish towns and cities.

The Irish Policy Context

Unlike in the United Kingdom where a distinct urban policy has been evident since the late 1960s and a brownfield policy evident from the 1990s, it is only in recent years in response to rapid and unprecedented urbanisation that urban management has become a critical issue in the Republic of Ireland. In 1997, in response to commitments made at the Rio Conference five years earlier, the government published *Sustainable development: A strategy for Ireland* (DoE, 1997) without a chapter on sustainable urban development. However, the promotion of sustainable urban development is aspired to in generic terms, focusing on the effective utilisation of existing developed urban areas and signalling a role for brownfield regeneration. Yet no attempt has been made at national level to devise an effective land use and regulatory policy that would assist local authorities in engaging in land recycling outside specially designated areas.

However this is beginning to change in response to the *National Spatial Strategy 2002-2020: People, places and potential* (DoE, 2002). This latest document provides a framework within which balanced and sustainable regional development could take place in Ireland over the next two decades, and argues for a strengthening of existing policies of counter-urbanisation to halt the further development of a large urban conurbation on the eastern seaboard. A policy of containment is called for and the promotion of effective land use and public transport policies and the potential for brownfields to contribute to intensification and compaction is made explicit. The document argues that within the Dublin Metropolitan area 'a systematic and comprehensive audit of all vacant, derelict and underused land [should be undertaken] to establish its capacity to accommodate housing and other suitable uses'. A practical prescription to ensure consolidation of the Greater Dublin Area and other large urban centres within Ireland is given: 'the efficient use of land by consolidating existing settlements, focusing in particular on development capacity within central urban areas through re-use of under-utilised land and buildings as a priority'. This suggests that brownfield regeneration has the potential to become a cornerstone of future spatial development, but much work remains to be done to ensure the opportunities are maximised. Given the manner in which many key provisions of the National Spatial Strategy appear to have already been undermined for political expediency, it is questionable whether brownfield regeneration will be encouraged, given the scale of the challenge that this poses. However the publication of updated Regional Planning guidelines for the Greater Dublin Area in July 2004 and the attempt by Dublin City Council in the Draft Development Plan to address this issue suggest that public servants have the will to make the change if the necessary supports are put in place.

Problems of Brownfields in Ireland

Three key obstacles remain to make the job of those who wish to effect change more difficult, and they require direct intervention by central government to ensure successful regeneration.

Their Extent

One of the most critical issues at present in Ireland is the lack of any formal research about the extent of brownfields, and the small number of individuals with limited resources trying to plug the information gap. There is currently no up-to-date and comprehensive survey in Ireland of the number of former industrial sites, although the Environmental Protection Agency (EPA) estimate that even after the large urban regeneration projects that were completed through the 1990s, between 2,000 and 2,400 potentially contaminated sites remain, mostly located in Dublin and Cork (CLARINET, 2002). These include former gasworks, closed mines, dockyards, landfills, fertiliser plants, railyards, old petrol depots and stations. This emphasis on contamination as a characteristic of 'brownfields' has resulted in the identification of 487 sites previously used for hazardous activities in the Republic of Ireland through the National Hazardous Waste Management Plan, and gives some indication of the potential effect that brownfields may have on ground and surface water resources. This number is not large in comparison to other countries, but until the figures compiled by Brogan et al. (1999) are up-dated and clarified the extent of the problem and the potential that these areas may provide for environmental restoration will remain unknown. One of the primary difficulties is that this issue falls between many stools, encompassing a broad range of aspects from waste management, to groundwater protection, pollution control, planning and development and both human and ecological health. Policy integration rather than a sectoral approach is critical to providing a comprehensive framewok for redevelopment, but given the policy ownership boundaries that exist between government departments in Ireland, it will require a centralised initiative to facilitate inter-agency co-operation.

Weak Policy

As earlier indicated, the introduction in the late-1990s of clear policies addressing the legal, political and financial aspects of brownfield remediation in the US and UK resulted in a significant increase in interest from potential investors. In Ireland, the primary legislation dealing with brownfield sites is the Derelict Sites Act introduced to address the issue of urban blight in urban cores. This legislation compels local authorities in urban areas to keep a register of all derelict sites, including the location of the property, the name and address of the owner and details of any action that the local authority has taken regarding the site. If the owner of a site does not engage in improvement, the local authority is empowered to fine them 3 per cent of the market value of the site per annum until such time as the necessary changes are made. The legislation also grants the local authorities power to buy derelict sites or dangerous land in their areas, either by agreement with the owner or compulsorily. However, this legislation has proved highly ineffective as it has only provided a framework within which local authorities may act without compelling them to do so. In some instances, private landowners have avoided penalties under this Act by erecting a palisade fence and thereby 'tidying' the site, even though it may remain vacant or under-utilised. Between 1993 and 1999, fines under this legislation yielded Dublin Corporation (now Dublin City Council) IR£90,000. However, as the local authority did not force acquiescence with the legislation, over IR£70,000 was left unpaid by private landowners (McDonald, 1998). The punitive nature of the legislation has been totally ineffective due to lack of enforcement and the management of the Derelict Sites Register. Many of the large and more obviously derelict sites throughout the city have not been entered on this list because of the poor procedures utilised to manage the database. It is also totally ineffective in determining whether a site is contaminated, as there is no indication of the characteristics of that place. Organisations such as the Dublin Docklands Development Authority (2003, p. 58) have criticised the narrow definition it adopts which 'does not provide an accurate indication of the widespread nature of under-utilised sites'. Although the National Spatial Strategy has compelled local authorities to use this legislation to solve the problem of brownfields and to undertake a comprehensive and rigorous audit of derelict land, more resources should be allocated to local government to undertake this analysis and prioritise brownfields as future growth centres. Dublin City Council, with minimal resources, has already identified 27.7 hecates of brownfield in the north inner city and rezoned it in an attempt to encourage redevelopment, but by their own admission the lack of resources has meant that their approach is neither entirely comprehensive nor totally accurate.

Liability

Perhaps the greatest concern from the developers' perspective is the lack of clarity regarding liability for derelict, and particularly contaminated, sites. Internationally this has been a major stumbling block to regeneration as a number of questions, including who pays for the clean up of contaminated sites, how clean is clean, and who is responsible for these sites once they have been de-contaminated, are left unanswered. Whether the polluter or new owner bears responsibility for any future liability from historical contamination is a critical issue, particularly given that the national media and politicians are increasingly referring to the emergence of a litigation culture in Ireland. Potential investors in brownfield areas may be unwilling to accept the risk that they could become liable for future compensation claims, and therefore the public sector may need to intervene with protection mechanisms similar to the US Superfund programme. Beyond the national level, European Commission directives and guidelines are of little use in this regard given that they do not cover contamination retrospectively. Apart from the end user taking a claim against a particular developer, there is also the employer liability that may arise from the remediation of a contaminated site. McIntyre (2003) argues that Section 2 of the Health, Safety and Welfare at Work Act, 1989 could potentially include a development site where personal injury claims may emerge. There are also constitutional difficulties (McIntyre, 2003). The Irish Parliament cannot declare an act to be an infringement of law retrospectively. Therefore if a landowner created a brownfield but the action was not illegal at the time, then they bear no responsibility for the contamination. It is therefore virtually impossible to legally assign liability for historically contaminated land in Ireland. It would therefore seems that because of legal and constitutional difficulties in this jurisdiction, 'responsibility [for regulating liability] is ultimately likely to pass to the public sector, perhaps explaining the Irish authorities' lack of resolve to tackle this problem' (McIntyre, 2003, p. 117). As in the United States, some form of state indemnification against liability from historically contaminated sites must be introduced.

Overcoming the Challenges

Despite the obstacles, what has already been achieved in other countries and indeed on a very small scale by Dublin City Council suggests that brownfield regeneration has the potential to become a cornerstone of future spatial development in Ireland, but a significant amount of work remains to be done to ensure that opportunities are maximised. Within this context and indeed leading the field in this area has been the Dublin Docklands Development Authority (DDDA) who have tackled and overcome the development constraints of one of the most strategically sited brownfields in Ireland, at the Grand Canal Dock.

Unlike much of the docklands area, which is relatively cut off from the rest of the city, the Grand Canal Dock is highly accessible and is familiar to anyone crossing the city regularly by car or train. For many years, the gasometers that dotted the skyline in this area could be spotted from quite a distance, setting this area within the city apart from any other (Figure 2.5). In addition, this area had a long history of other noxious industries including chemical works, coke works and coal yards, forming one of the two key industrial zones in nineteenth century Dublin.

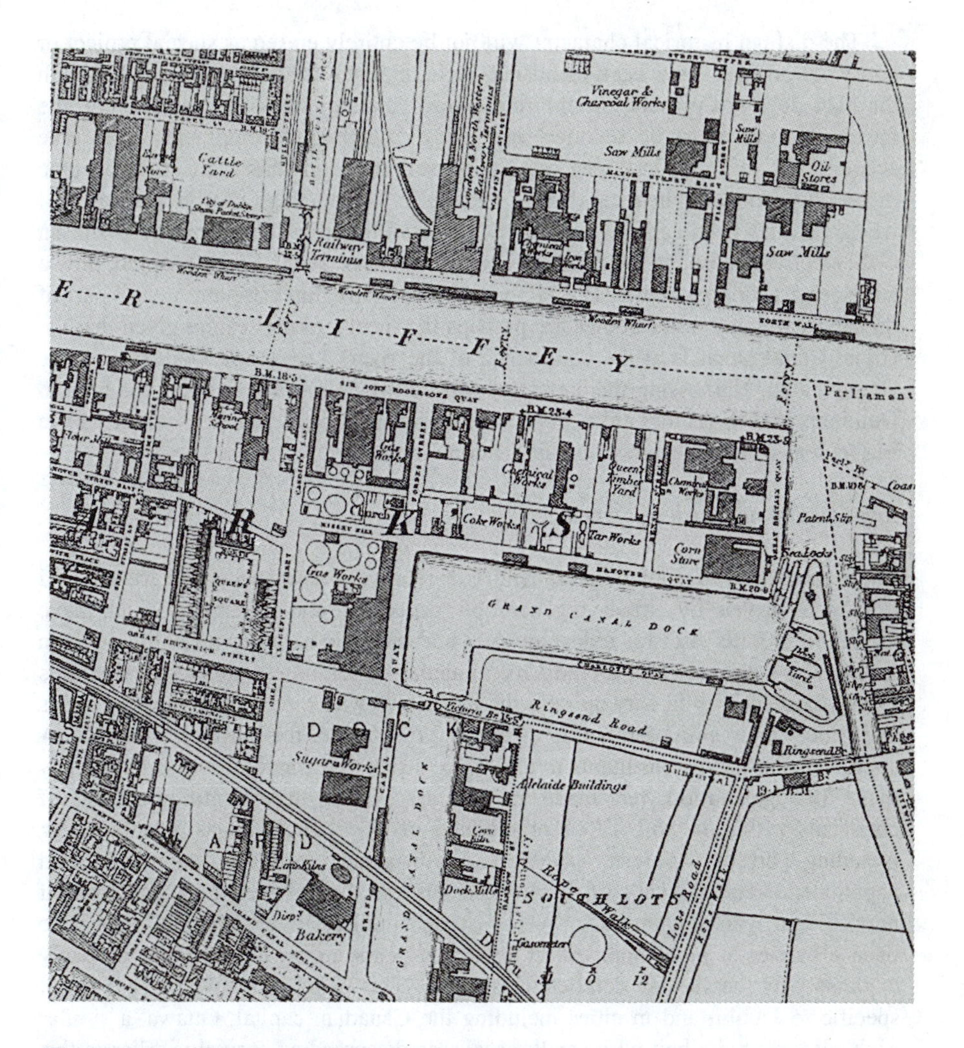

Figure 2.5 The Grand Canal Dock, 1876

The development anticipated by the Dublin Docklands Development Authority for this area, and currently under construction, closely resembles the original 1987 Planning Scheme for the Custom House Docks Area, as it aspires to the creation of a vibrant mixed-use development with 40 per cent residential land use, comprising dwellings, cultural, hotel and local retail uses, and 60 per cent commercial land use.

However, given the slow down in the office market and the need to breed vibrancy and promote the 'people factor' in this district, the balance has been reversed and the project will now become 60 per cent residential, creating an entirely new living quarter in an area that has always been associated withindustrial activity and mirroring the success of similar provisions in the United Kingdom.

The distinct historical character will not be entirely erased as special project or conservation areas have been identified, including two listed warehouses. Although the plan does not provide for the retention of the structures in their entirety, the façade of both must be retained and the surrounding developments must take account of this. Undertaking urban conservation in this way has become increasingly popular in recent years in Dublin. Recent high-profile examples of where this has been achieved relatively successfully are at the former AIB site on College Green, now the Westin Hotel, and the former Jervis Street hospital, now a large-scale shopping centre. But although historic structures are an integral part of the landscape, the water bodies are perhaps the most striking feature given that the Grand Canal Dock is at the confluence of the rivers Liffey and Dodder, and the Grand Canal. Harnessing this potential will be important in promoting the area to Dubliners and to visitors alike, but before this can be delivered the Authority have had to deal with the legacy of industrial contamination.

Remediation and Redevelopment

The remediation of contaminated land is a relatively new concept in Ireland and has been driven by urban regeneration processes and agencies, which have identified specific sites for redevelopment and then had to deal with pollution from former use activities. As already mentioned, the Grand Canal Dock Area was synonymous in Dublin City up until the late 1970s with the production of town gas and since then with dereliction (Figure 2.6). Given this history, the DDDA appointed Parkman Consultants to undertake a risk assessment exercise before any plans for regeneration were made. In their assessment, they identified a range of chemicals within the soil, all either products or by-products of gas manufacturing including cresol, toluene, xylene, ethylbenzene, benzene, ammonia and naphthalene. The presence of these chemicals and the knowledge gap in terms of how to deal with contamination which existed until relatively recently, may have been a barrier to redevelopment at what is a prime urban location and probably explains why the site lay derelict for so long. This delay in redevelopment is not specific to Dublin and in cities including the Canadian capital, Ottawa, a similar problem emerged albeit on a smaller scale, as discussed extensively earlier in this chapter.

The IBI Group had argued for dumping the waste in the Ottawa river pushing the problem elsewhere. However at the Grand Canal Docks in Dublin, this was not an option given the strict environmental regulations governing the clean-up of these sites and the attitude of the Dublin Docklands Development Authority which Peter Coyne, Chief Executive of the DDDA, sums up as 'a belief that you must spend money to make it', a philosophy evident from the eventual €50 million bill for the remediation of the 24 acre site at Grand Canal Harbour. In order to begin construction and satisfy human and environment health concerns, this site has been certified by the Environmental Protection Agency, which played a monitoring role during the remediation process. The EPA licensed two

Figure 2.6 Grand Canal Dock prior to redevelopment
Source: DDDA

separate development sites, both of which abut Sir John Rogerson's Quay within the Grand Canal Harbour boundaries, in October 1999. The first part of remediation involved the construction of an 8m deep and 2km long bentonite wall into the soil to ensure that any contamination on the site or any dislocation caused by remediation would not pollute adjacent areas. The entire site was then excavated to a depth of four metres, and the groundwater, which may have been contaminated, was pumped out to a sewer via treatment works. Hazardous soils were shipped off-site to a number of EU countries, including the Netherlands which burns the soil clean and then utilises it in the construction of flood defences, but most soil was treated on site, as the EPA prefer, and then reused.

The Waste Licence for the first part of the site was surrendered in December 2001 and the second one a year later, following the removal of over 134,000 tonnes of material and a final inspection by the EPA. This achievement was a key victory for the DDDA who had demonstrated that such risks could be overcome, and it bodes well for future development in the country as a whole given that the EPA estimate there to be about 2,400 brownfield sites in Ireland, although none as large as this one. Beyond the physical remediation, one of the most difficult challenges to address during the whole process was public perception. However in an important public relations exercise and to address and curb community concerns, fortnightly public meetings were held by the DDDA to report on progress. Now the Authority argue that the air around this site is among the cleanest in the city, and is certainly cleaner than College Green which has some of the largest pedestrian flows in the city. An additional investment of €7 million for infrastructure and other environmental improvements was made by the DDDA, but this has been easily justified given that the Authority has already generated over €200m in land sales from this site.

A range of tenants have been secured for Grand Canal Harbour, the 25 acre site being developed directly by the DDDA within the overall 92 acre dock area (Figure 2.7). These include the headquarters of legal firm McCann Fitzgerald, who will occupy Riverside 1, adjacent to the Ferryman hotel. This site was originally designated for the software firm, Novell, who proposed to pay €25.4m for the site and construct a 6/7-storey tower due for completion in 2002. However the collapse of the technological sector in recent years precluded this occurring and McCann Fitzgerald acquired the site for a bargain €12.5 million. When the building is complete, they will relocate from Harbourmaster 2 within the IFSC, but must find another tenant to take up the lease there given that it does not expire until 2026. This marks a worrying trend in relation to the overall docklands area. Given its size and the likelihood that newer and better office facilities will continue to be constructed over the coming ten years, a detrimental impact may be felt in the original Custom House Docks site where tax incentives have now ended. If firms identify a preferable location elsewhere in docklands, the prestigious nature of the original IFSC buildings may be undermined.

As well as the commercial element, which is after all not the majority land use within the Grand Canal Dock, the strength of the 'market' for a range of other uses appears to be strong and the Park Hyatt Hotel Group have just recently confirmed that they will be constructing a 5 star hotel at Misery Hill and an adjacent Hyatt Residence Hotel. By the time the development is complete, there will be over 3,500 new apartments in the area and over 4.4m square feet of office space. Five 20-storey landmark towers will punctuate the site, the most well-known being the new U2 tower which will house a recording studio for the band. The whole development will be anchored by a 1.5 acre piazza, Grand Canal Square, adjacent to which will be a performing arts centre, designed by the internationally renowned architect, Daniel Liebskind, and developed by Heritage Properties. In complete contrast to the attitude adopted towards Stack A and similar to the incentive used to attract the National College of Ireland to docklands, part of the site for this new cultural amenity has been provided free of charge. A development levy applied to all commercial tenants should raise approximately €500,000 to subsidise the centre, which it is anticipated will encourage a wide range of visitors to the site outside the traditional working hours.

What the Development Authority has succeeded in doing is engineering a dramatic transformation in the desirability of this former, marginal part of the city centre, and the result has been massive spin-off benefits for landholders in the immediate vicinity. New projects outside the Grand Canal Harbour boundaries

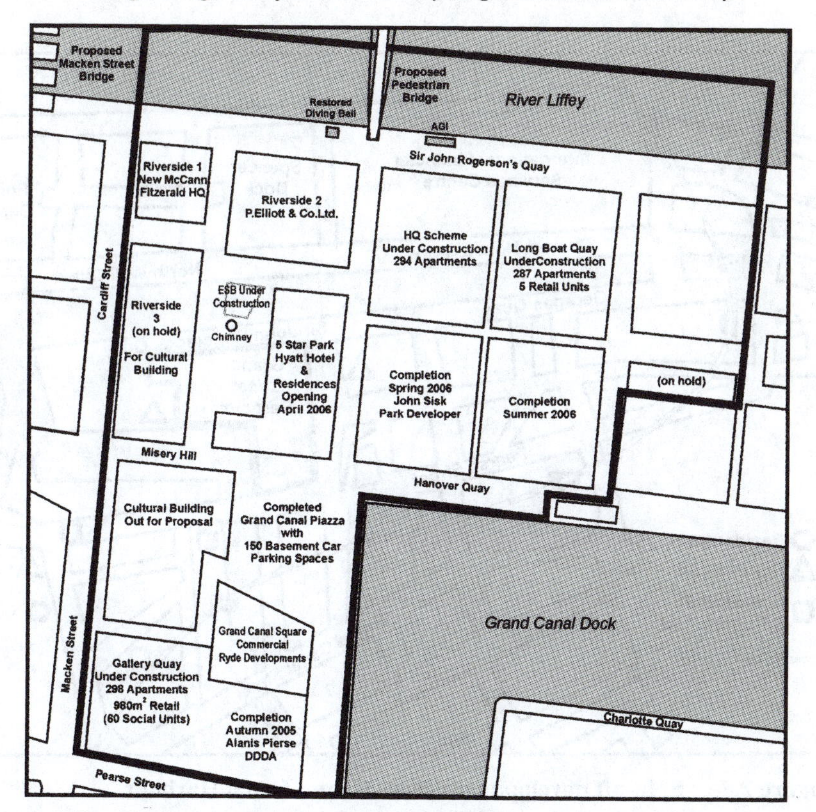

Figure 2.7 Tenants at the Grand Canal Harbour redevelopment

include the Duffy Lally development at South Bank Quay and the Ellier Developments project at Hanover Quay. Treasury Holdings have just completed contracts on 230,000 square feet of office space fronting Sir John Rogersons Quay, while another joint venture development with CIÉ will create a mixed office-apartment scheme near the Grand Canal Dock DART station (Figure 2.8). If the mixed-use development is successful, this may prove to be a significant turning point in terms of how the potential of brownfield areas is perceived, and signal a new trend in Irish planning and development away from greenfield sites and back into these residual, city-centre 'problem areas'. The idea that developers can make significant profits from such sites, outrageous in an Irish context four years ago, now quite remarkably, has more than a ring of truth about it.

Conclusion

Unlike Ireland's record in other environmental areas, the record on recognising the problems and potential of brownfield redevelopment is poor. In 1997, the Irish government led the way in dealing with water management introducing a new

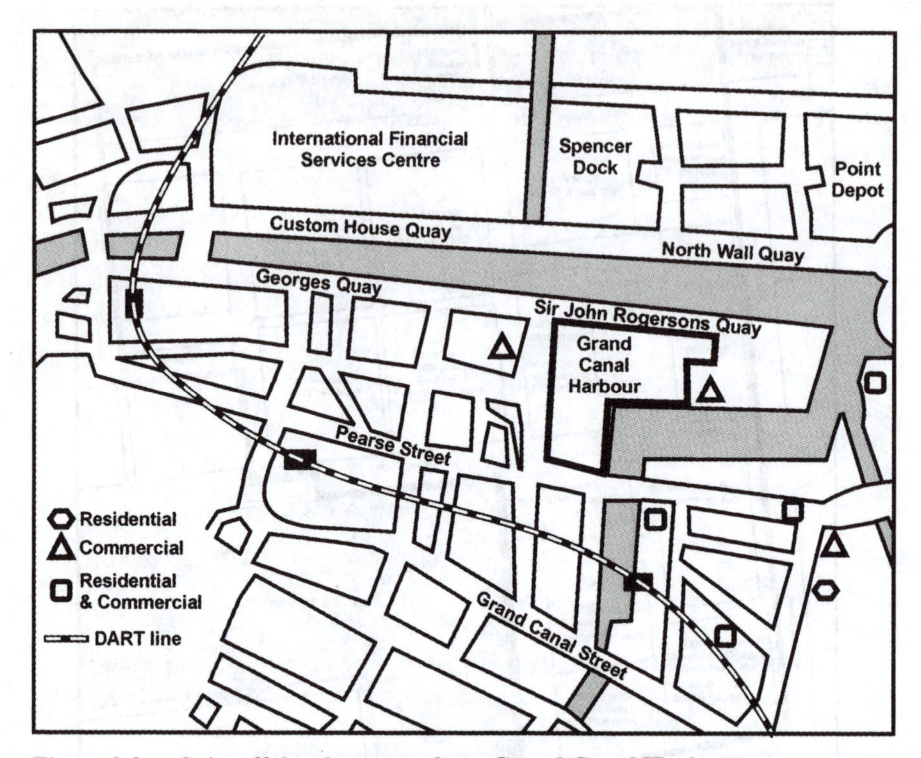

Figure 2.8 Spin-off developments from Grand Canal Harbour

approach to basin management, pre-empting the EU Water Framework Directive of 2000. Given that a similar integrated approach to land management does not exist, the opportunities provided by brownfield regeneration to turn aspirations into practical realities will be missed if new policy options are not considered.

Critical to realising these goals is the adoption of an integrated policy approach, co-ordinated across geographical and sectoral boundaries and flexible enough to cope with changing circumstances. The idea that brownfield regeneration is simply an urban problem must also be eradicated and the broad benefits for all parts of the country, including more efficient and viable public transport systems to the protection of agricultural land, must be promoted.

Drawing on the experience of the UK, a sequential approach to development favouring brownfield over greenfield sites for new development could be adopted or at least debated. A similar policy to that introduced in Ottawa, Ontario in 1995, which stated that extensions to the settlement area would only be permitted if 'the amount of land included within the extensions is justified, based on the amount of land available for development in the settlement area, and on population projections and employment targets for the municipality for a horizon of 15-20 years', could have immediate benefits, particularly if coupled with better financial support to local authorities for environmental restoration projects and forced compliance with regulations for developers. This 'stick' approach should be provided in conjunction with incentives to developers, which could include the allocation of tax incentives over the next five-years only to brownfield sites and the imposition of a greenfield tax if a sequential approach to development is shunned.

What the Grand Canal Dock project has illustrated is that where there is a will and central government support to achieve a goal, it is possible and can in fact result in economic and social returns far beyond the initial investment. It is time that Irish government policy, rather than paying lip-service to the issue of urban sustainability, seriously heed the conclusions of the OECD (1996, pp. 14-17) that:

> Cities cannot solve environmental problems when national policies are incoherent, or leave in place fiscal or regulatory instruments that promote environmental degradation, or deny cities the technical and financial resources they need ... An ecological city is distinguished by the degree to which environmental considerations are incorporated into decision-making in public and private sectors alike.

References

Bjelland, M. (2000) 'Brownfield sites: Causes, effects and solutions', *Centre for Urban and Regional Affairs Reporter*, University of Minnesota, pp. 1-9.

Brogan, J. et al. (1999) 'Ireland'. In Ferguson, C.C. & Kasamas, H. (ed.) *Risk Assessment for Contaminated Sites in Europe,* Vol. 2, Policy Frameworks, Nottingham: LQM Press.

CLARINET (2002) *Brownfields and Redevelopment of Urban Areas: A report from the Contaminated Land Rehabilitation Network for Environmental Technologies,* Austria: Federal Environment Agency.

Comhar (2002) *Principles for Sustainable Development,* Dublin: Stationery Office.

DoE (1997) *Sustainable Development: A Strategy for Ireland*, Dublin: Stationery Office.

DoE (2002) *The National Spatial Strategy, 2002-2020: People, places and potential*, Dublin: Stationery Office.

Doak, M. et al. (2003) 'The remediation of contaminated land in the Republic of Ireland', *Proceedings Sardinia 2003, Ninth International Waste Management and Landfill Symposium,* Cagliari, Italy 6-10 October, 2003.

Dublin Docklands Development Authority (2003) *Dublin Docklands Area Master Plan*, Dublin: DDDA.

European Sustainable Cities Project (1996) *Progress report, 1993-1996,* Brussels: European Commission.

European Union Expert Group on the Urban Environment (2001) *Towards more sustainable urban land use,* Brussels: Eurocities.

Grimski, D and Ferber, U. (2001) 'Urban brownfields in Europe', *Land Contamination and reclamation*, Vol. 9(1), pp. 143 – 148.

Kelly, M. (unknown) *Evaluating Sustainable performance*, Research paper, Environmental Protection Agency.

McDonald, F. (1998) 'Derelict Sites', *Property Valuer*, Autumn 1998.

McIntyre, O. (2003) 'Problems of liability for historical land contamination under Irish law', *Irish Planning and Environmental Law Journal*, Vol. 10(4), pp. 112-118.

OECD (1996) *Innovative policies for sustainable urban development: The ecological city*, Paris: OECD.

Satterthwaite, D. (1997) 'Sustainable Cities or Cities that contribute to sustainable development', *Urban Studies,* Vol. 34(10), pp. 1667-1691.

Syms, P. (1999) 'Redeveloping brownfield land: The decision-making process', *Journal of Property Investment and Finance*, Vol. 17(5), pp. 481-500.

Chapter 3

Higher Density or Open Space: The Future Role of Green Space

Craig Bullock

Introduction: Green Space and the Built Environment

Recent economic growth has finally allowed Ireland to begin to catch up with its European partners in terms of an array of quality of life indicators such as per capita income. In the meantime this same growth has placed pressure on the existing infrastructure. While true of the whole country, infrastructure constraints are certainly very evident in Dublin, most recognisably in the familiar level of road congestion. Painful as it may now seem, State and Structural Fund investment will eventually see the infrastructure deficit overcome. However, if quality of life indicators are to have any meaning in the future, it is essential that policy-makers and planners give thought to the type of urban environment that will justify this investment in the long-term.

This chapter deals with one aspect of that environment, namely parks and other areas of urban green space. It is generally acknowledged that green space provides many benefits in terms of amenity, direct recreation opportunities, health, child development, social interaction and community identity. In addition, green space has the capacity to provide other benefits such as habitat for wildlife, dust filtration, noise mitigation, climate moderation (wind and temperature) and flood control.

The problem is that it is very difficult to measure these benefits. Pollution and climate benefits can be measured through a mixture of physical changes and economic values, although this would still make for a challenging cost-benefit analysis. The benefits in terms of amenity or community well-being are yet harder to identify. Yet green space is not a costless resource. It requires maintenance and also sterilises valuable land from profitable residential or commercial development. It therefore makes sense to quantify the benefits in economic terms if at all possible so that these can be measured against the costs. In the absence of such information, green space management has been a low priority for government and local authorities faced with competing resource demands from the likes of health, roads and education.

Recognition of the overall value of green space is not helped by the fact that it is also a public good in the sense that it is not privately owned and is freely accessible. As a consequence, while people might individually value certain green spaces, rather few turn up for meetings where the public have been invited to participate in the management or maintenance of local green space. The exception is where there is an element of quasi-ownership as, for instance, where areas of green space are largely used by a single small community. When people purchase a home they generally seem to be willing to trade off shared community assets for private benefits such as a larger area of floor space (Rowley, 1996). Indeed, respondents to surveys have expressed a greater belief in the contribution of private garden space to their property value than in that of adjacent areas of pleasant communal green space (Peiser and Schwann, 1993). Nevertheless, when asked what it is that they most value about the community in which they live, people often mention green space. The much maligned English town of Milton Keynes has many adherents amongst its residents for precisely this reason (Finnegan 1998, Allen et al., 1998).

This is not to suggest, however, that planners have failed to recognize the benefits of green space. Green space provision has been an integral part of modern town planning since its inception. In nineteenth century America, the 'Picturesque movement' elevated wilderness preservation to an almost spiritual dimension as the 'visual antithesis to the gridded streets' of the city (Cranz, 1989). Closer to home, Ebenezer Howard's Garden City Movement promoted the concept of a self-sufficient community surrounded by market gardens and green space. In fact, the community of Marino in Dublin, designed by Raymond Unwin in the early part of the twentieth century, represents one of few planned working class communities that have been influenced by the Garden City idea. Generally, though, it was the middle classes that bought into the concept, but rather in the form of leafy new towns and suburbs graced with ample parkland.

This suburban ideal is well-rooted in Ireland, Britain and, of course, North America. The expansion of the suburbs has presented numerous opportunities to purchase large country estates, or demesnes, in return for development rights. Often these estates have been preserved as parks and many, such as Marlay Park or Bushy Park in South Dublin, are extremely popular. The acquisition of green space has also been reinforced by quantitative standards and official advice that a proportion of open space must be retained in any new residential development. These standards have been quite generous. In Dublin, it is recommended that at least 10 per cent of any new development is set aside for green space.

There is no merit in ignoring the popularity of the suburbs. People associate the middle-class, semi-detached, low density suburbs of South Dublin, Dun Laoghaire and Fingal with personal success and advancement. The same blueprint has applied as the urban periphery has extended into the hitherto rural counties of Meath and Kildare. People have voted with their feet. They have voluntarily moved out to the suburbs in recognition of the better schools, greater space and greenery that they provide. Recently, these pull factors have been joined by the push factor of locating affordable housing.

The low-density suburban model is now often scorned by planners if not, of course, by many suburban inhabitants or developers. The suburbs have become associated with sprawl, car dependency, and social isolation. The problem is that, while people resent the traffic jams they experience when travelling from the suburbs into town, they believe, with much justification, that the solution lies in improved roads and transit systems. There are currently few examples of alternative residential development to which they could aspire.

The huge inconvenience of traffic congestion does, however, provide an opportunity to promote alternatives. There are opportunities to provide new higher density development of the type described in the Residential Density Guidelines (Department of Environment and Local Government, 1999). In those parts of the urban centre that have been targeted for urban regeneration and renewal, the concept of an urban realm has been contemplated. This would involve open space being arranged together with pedestrian walkways so as to provide people with access to parks and community facilities without the need to resort to the car. Compact squares, landscaped linear corridors and greenways are all part of this vision.

However, urban redevelopment rarely occurs on a scale to facilitate these laudable ideas. There can also be a lack of conviction to transfer such concepts from drawing board to reality. In part, this has been due to a rather unequal culture of partnership with private developers (McGuirk, 2000). Although planners' willingness to assert grand ideas has been growing in line with the investment opportunities presented by the new economic environment. Nevertheless, there are still many lamentable examples of isolated island apartment development in suburbs such as Kilmainham. Such development does little to promote the idea of higher density development to those looking to climb onto the property ladder.

Mixed-use residential development has been proposed, for example in the design for Adamstown in South Dublin. This concept has sought to overcome the isolated character of much higher density apartment development. Much depends on the planned transport infrastructure being realized, but it is hoped that such development will appeal to a wider demographic cross-section than has been the case for higher density housing to date. It has yet to be seen if this model will win over those with dependent children for whom private gardens have hitherto been a necessity.

The Role of Green Space

In the Residential Density Guidelines, much is made of the need to provide high quality community facilities if people are to be converted from the suburban model to higher density living. Amongst these better quality facilities, the Guidelines make reference to green space. As the suburbs currently contain much pleasant if rather bland green space, it is important that new green space provision is of a better quality than has typically been experienced by most Dubliners to date.

But what is meant by 'better quality'. It is hard to imagine most people being able to provide a definition without making reference to the obvious examples of Dublin's more attractive parks. These examples would include Merrion Square, Marlay Park, Bushy Park, St. Anne's Park and, for some people, the more natural expanse of meadows and trees to be found in Phoenix Park. There are few actual examples of anything resembling the green elements of the urban realm. Likewise, there are few examples of green space which has been managed for wildlife, for heritage appreciation or for the regular holding of public events. There are some examples of linear greenways such as those along the rivers Dodder, Tolka or Santry. Indeed, a network of greenways following the city's watercourses and linking the main existing parks, had been proposed for Dublin by the planner Patrick Abercrombie as far back as the beginning of the twentieth century. Unfortunately, while there are pleasant stretches, for instance, in Rathgar and Clonskeagh, much of the remaining greenway is discontinuous, featureless or poorly maintained.

In fact, aside from a handful of good examples, much of Dublin's green space can most politely be described as being rather dull. Much of that which is to be found in north Dublin, for instance, Finglas or Coolock, is completely featureless and has been actually described as 'prairie'[1] (except that the prairies would contain more interesting flora!). Perhaps not surprisingly, while our GREENSPACE[2] survey found that 45 per cent of people 'frequently' or 'quite frequently' use parks, there are an additional 33 per cent who 'rarely or never' visit parks.

Back in the 1960s, Jane Jacobs, in her critique of American town planning (Jacobs, 1964), described much of the country's urban parks as being monotonous, as though they had been 'rolled with a dye-stamper'. The solution, she proposed, was to provide diversity and to locate parks in such places to attract use throughout the day by all social and demographic groups. Evidently, her criticism could be applied to Dublin's parks forty years on. It is a good question as to whether the proportion who 'rarely or never' use parks would remain so high if the city's parks were more interesting and diverse.

Unfortunately, following the elimination of domestic rates, local authorities have become largely dependent on meagre central funding which has to be allocated to such priorities as roads, schools, policing and waste management. In comparison, green space is considered a luxury by county managers. Limited resources have meant that parks have been subjected to a homogeneous management regime that favours scattered trees and mown grass. This type of management has been reinforced by concerns with liability and safety that have led to the removal of the shrubs that provide home for wildlife. For much the same reasons, there are lamentably few good playgrounds to be found in Dublin parks. Most of these are of a standard metal design. Yet families are amongst the most frequent users of parks.

The GREENSPACE Survey

To investigate what types, or portfolio, of green space would be most valued by Dubliners, the GREENSPACE study asked people about their fundamental needs in relation to green space. The study set about this task in three ways. Firstly, focus groups were used to help determine what people most enjoy or dislike about green space. Secondly, a factor analysis survey was undertaken using both a postal questionnaire and in-park interviews to determine how people rated various attributes of green spaces. Finally, as the largest component of the study, 500 door-to-door interviews were undertaken incorporating a choice experiment method. People were presented with a set of eight 'choice sets' in which they were asked to choose between two alternative park descriptions, and then to compare these alternatives with their usual park destination.

The Focus Groups

In the focus groups, participants are encouraged to discuss particular topics with minimal intervention from a facilitator. The technique has often be used to provide preliminary information prior to undertaking survey work, but also represents a form of social inquiry in its own right (Kitzinger, 1994). Much of the benefit of the technique derives from the group interaction. Focus groups cannot be considered representative as they comprise only a handful of individuals, but they can provide evidence of underlying beliefs or motivations. Often focus groups have been used to shed light on sensitive subjects, although, for green space, their benefit is rather in providing people with a forum to debate a topic that, while valued, may be rarely discussed on a day-to-day basis.

Discussion groups have been used before to investigate green space use, notably by Burgess et al. (1988) in the UK. Amongst other findings, their study reported on the value of informal spaces. Participants recalled their childhood enthusiasm for wasteland and other informal open spaces and lamented the loss of access to such places due to current preoccupation with crime. Similar comments were made by the participants of the three focus groups held by GREENSPACE. Natural areas of green space were indeed valued, but participants needed to be reassured by evidence of security and active maintenance. Facilities such as seating, paths, toilets and places to eat were also regarded as important. Indeed, facilities, safety and maintenance (litter collection, repairs, pond cleaning, etc.) were perceived to be the most important pre-requisites for green space use.

The Factor Analysis

Many of the same results emerged from the factor analysis. In this component of the survey, people were asked to rate over thirty different green space attributes on a seven point scale from 'like very much' to 'dislike very much'. Amongst the most highly rates attributes were 'frequent removal of litter', 'adds to reputation of community', 'good for other people in the community' and 'has a park keeper'. The most disliked characteristics included 'discarded glass and bottles', 'problems of litter' and 'popular with teenagers'. However, there was a wide variation in opinion towards some attributes such as woodlands.

However, while rating is interesting in itself, it is hardly a sophisticated process. The more interesting output from factor analysis is its ability to identify correlations between the ratings of different attributes. Through a process of rotation, these attributes can be grouped into key components. The principal component to emerge included a mixture of variables that together could be described as a 'peaceful, safe community amenity', namely a typical local park with good facilities such as paths, seating and playgrounds. The identification of such a construct is not unsurprising, particularly given the output from the focus groups. More interesting, though, were the second and third components. The first component, 'naturalness', revealed that many respondents value attributes that together describe a natural setting (this value appeared to be associated as much with naturalness itself, than the opportunity to see wildlife). Interestingly, this preference did not extend to all forms of low intensity maintenance such as 'irregular cutting back of shrubs and undergrowth'.

Similarly, there was an interest in opportunities for active recreation, although more in the form of long walks, trails and cycle routes, rather than the playing fields that dominate many parks. As with naturalness, this preference was not universal, and would otherwise be disguised behind the average ratings.

Choice Experiments

Economists generally feel uncomfortable with the process of ranking or rating. They prefer analysis based on choice given that real consumption decisions are not founded on expressions of relative preference. In the choice experiments, respondents were asked to choose between two alternative parks described by their component characteristics. These attributes were presented at various levels. For the example of vegetation, these levels were 'mostly mown grass, few trees', 'parkland with scattered trees' and 'mostly woods and meadows'.

The attribute levels were combined within an experimental design in which every level is combined with every other level so that unwanted correlation is avoided. In this way, a combination of attributes within a choice experiments can together describe a particular type of green space. If, instead, analysis were to be based on actual use, i.e. revealed preference, the combination of attributes would be restricted only to those that already exist. Furthermore, the fixed nature of many attribute combinations would obscure the value of any one attribute considered in isolation.

In total the exercise was restricted to seven key green space attributes, namely size, maintenance, vegetation, water features, play facilities, seating, paths and trails, number of users and journey time. The choices that individuals make between combinations of these attributes could be modelled so as to provide a quantitative estimate of the marginal value of any one attribute level. In addition, because Journey time has an opportunity cost in terms of alternative income-earning activity, it was possible to quantify the value of each attribute level in monetary terms for the purposes of a cost-benefit analysis.

Model fit was restricted by the limited number of attributes that could be presented. However, the large number of attribute combinations allowed an examination of the interactions that exist between different attributes and between the valuations of these attributes by certain population subsets. The good significance of estimated parameters suggests that most coefficients are reliable for these attribute levels.

Survey Results

The analysis provides coefficients for each of the attributes in relation to the base level. For a three level attribute, such as Trees, the base level is 'mostly mown grass, few trees', i.e. Tree1. The coefficients indicate the probability of selecting the chosen alternative combination given the presence of a particular attribute level.

The results reveal that distance, or Journey Time, is amongst the most significant variables. The variable is negative, demonstrating that there is a disutility associated with increased journey time. Furthermore, the significance of the variable is increased if it is squared. This suggests that the disutility increases with greater distance.

The experimental design allowed for a relationship between park size and distance that would likely occur in reality, namely that small local parks are often just five minutes walk away, while large parks occur at a greater distance. Where this relationship is accepted, the influence of distance is such that small local parks are preferred. If the size/distance relationship is not accounted for, then larger park size becomes a positive attribute that increases the chances of that park being chosen in the choice experiment.

Amongst the other attributes, good facilities (Facs2) in the form of 'plenty of seating and footpaths' has the largest coefficient. The coefficient on a higher level of facilities (Facs3) including trails and cycle paths is actually slightly less, though still positive. The earlier focus groups and factor analysis had indeed indicated that the inclusion of cycle paths would not be to everyone's taste given conflicts between pedestrians and cyclists.

The presence of a 'mix of quiet and more busy areas' (People2) is the next strongest attribute, while in contrast, busy parks (People3) has a large negative coefficient, suggesting that the presence of many other people is a discouragement.

Play facilities have positive coefficients and this is highest for 'adventure playparks' (Play3). 'Scattered trees' (Tree2) have a positive influence on choice too, but this does not extend to 'woods and meadows' because Tree3 is negative. This negative coefficient possibly arises from concerns with personal security, although this same attribute level becomes slightly positive where the data is restricted to large parks only. Of other physical attributes, 'Riverside walks' (Water3) has a respectable positive coefficient, but that for 'natural-looking ponds and lakes' (Water2) is very small and not significant in this particular model. Once again, though, Water2 has positive value is the analysis is restricted to large parks.

Interestingly, more intensive maintenance (Maintain) has only a very small coefficient that is not significant. While the coefficients on Facilities are consistent with the focus group discussions, the participants' concern with good maintenance does not appear to be reflected in the model results. However, this result follows partly from the description of maintenance, whereby Maintain1 was described as less intensive maintenance rather than poor maintenance. Nevertheless, only the higher level of Maintain2 includes the reassuring presence of a park keeper. Table 3.1 gives the coefficients for each of the attribute levels and the respective monetary values based on the coefficient for Journey Time.

Table 3.1 Attribute coefficients and marginal economic values

Parameter	Coefficients	Attribute value per visit
Size	-.205	-€3.07
Maintain (high)	.013	€0.06
Trees3	-.197	-€0.95
Trees2	.140	€0.67
Trees1	-.057	€0.27
Water3	.071	€0.34
Water2	-.028	-€0.13
Water1	-.043	€0.21
Play3	.135	€0.65
Play2	.146	€0.70
Play1	-.281	-€1.33
Facs3	.288	€1.38
Facs2	.186	€0.89
Facs1	.474	-€2.27
People3	-.307	€1.47
People2	.197	€0.95
People1	-.110	€0.53
Journey Time	-.033	
Log-lik (parameters)	-1941.61	
Log-lik (no param)	-2196.13	
Adjusted ρ^2	.113	

Note: MNL model. Higher model fits available for different population.

Furthermore, it is possible to estimate the contribution to economic welfare of a change in a set of attributes. For example, it is possible to combine the marginal values for those attribute levels that are valued most by respondents on average. Where a comparison is made with the baseline level of each attribute (without subsets and with inclusion of interactions with reference to journey time), the respective compensating variation welfare gain amounts to €7.19 per trip. The estimates are a little insecure in that there are problems in assuming that the opportunity cost of leisure time can be extended to leisure journey time. If this assumption is accepted, a multiplication by the average number of trips made by respondents per year results in an annual value of €165. This figure can be considered a minimum given that park visits can be expected to increase if quality or variety were improved.

Varying Preferences

A restriction of the multinomial analysis of the choice data is that the coefficients represent average preferences. However, these averages can conceal the diversity in preference. This was evident in the factor analysis. As an alternative to the multinomial logit analysis above, mixed logit can be used to identify the sources of variation (or error) in the data. In many cases, this variation will relate to the respondents' socio-demographic characteristics. In other cases, the variation cannot be connected with a characteristic such as age or gender, but is more general throughout the population.

Where mixed logit analysis is applied to respondents with dependent children it is no surprise to find that the coefficients on play facilities are higher. Interestingly, that for 'natural-looking ponds and lakes' is negative, possibly due to parents' concern with safety. However, safety does not appear to be a consideration for woodlands. Neither are personal security concerns mitigated by numbers of people in the park. Personal security does, though, appear to be a factor in a negative relationship between female respondent choices and woodland (Trees3).

Further variation, or heterogeneity, in preferences is evident for 'age' where there is an unexpected positive association between older age and woodlands/meadows, but a negative association with busy parks (People3). Heterogeneity is also present for frequent and less frequent visitors. More frequent visitors appear to appreciate the presence of playground (Play2), a result that is consistent with the observation that people with children are amongst the most frequent visitors to parks.

Distance is also a key consideration for those belonging to lower socio-economic groups for whom the disutility associated with distance appears to be greater. As noted above, distance is always negative and significant. However, if the typical timing of a visit is accounted for, i.e. weekdays or weekends, the negative effect of distance is almost eliminated. This suggests that once people have committed themselves to a decision to visit a regional park, distance is no longer an important consideration.

Finally, mixed logit can be used to reduce this influence of status-quo bias where respondents were offered a choice between two hypothetical parks and their usual park. Without this approach, the model fit for responses to this question are good at a ρ 2 of .35, but the influence of occasions when respondents choose their usual park leads to suspect coefficients for some attributes. Some of these coefficients are indicative of the restricted attribute combinations that are currently available to people. For instance, busy parks (People3) has a positive coefficient in contrast to the hypothetical comparison. This reflects that fact that more attractive parks are indeed busy by virtue of their popularity.

By allowing for greater tendency for some respondents to opt for the status-quo, the mixed logit produces more accurate coefficients. Higher coefficients for riversides and play facilities are revealed. These attributes are often not available or are poorly provided for in most Dublin parks.

Conclusion

Although there are many attractive examples, much of Dublin's green space could be described as being rather homogeneous and bland. This type of green space is typical of the city's spacious suburbs and more of the same has been provided as the city has expanded in recent years. The chapter discussed how considerations of transport and sustainability have caused planners and urban designers to reassess the suburban model. Higher density development, including urban regeneration, has been proposed as a partial solution. However, there are currently rather few good examples of this alternative that could convert people from the suburban tradition.

Along with other factors such as quality architecture and streetscapes, diverse and attractive green space would need to be a part of any alternative model. This chapter has briefly described some of the results of the GREENSPACE project which demonstrate how relative values can be placed of some the key attributes of green space. These techniques can be applied to encourage planners and designers to provide alternative types of green space.

Notes

1. Colm McCarthy quoted by journalist Frank McDonald in his book *The Construction of Dublin* (2000).

2. GREENSPACE is an EU Framework 5 funded study of the benefits urban green space in six European states, www.green-space.org.

References

Allen, J., Massey, D., & Cochrane, A. (1998) *Rethinking the Region*, London: Routledge.

Burgess, J., Harrison, C. M., & Limb, M. (1988) 'People, parks and the urban green: A study of popular meanings and values for open spaces in the city', *Urban Studies*, Vol. 25, pp. 455-473.

Cranz, G. (1989) *The Politics of Park Design: A History of Urban Parks in America*, MIT Press.

Department of Environment and Local Government (1999) *Residential Density Guidelines (consultation draft of guidelines for Planning Authorities)*, Dublin: Stationery Office.

Finnegan, R. (1998) *Tales of the City: A study of Narrative and Urban Life*, Cambridge: Cambridge University Press.

Jacobs, J. (1964) *The death and life of great American cities*, Harmondsworth: Penguin.

Kitzinger, J. (1994) 'The methodology of focus groups: The importance of interaction between research participants', *Sociology of Health and Illness*, Vol. 16(1), pp. 103-121.

McGuirk, P. (2000) 'Power and policy networks in urban governance: Local government and property-led regeneration in Dublin', *Urban Studies*, Vol. 37(4), pp. 651-672.

Peiser, R. B. & Schwann, G. M. (1993) 'The private value of public open space within subdivisions', *Journal of Architectural and Planning Research*, Vol. 10(3), pp. 91-104.

Rowley, I. (1996) *Quality of Urban Design - A Study of the Involvement of Private Property Decision-makers in Urban Design*, London: Royal Institute of Chartered Surveyors.

Chapter 4

Suburbanising Dublin: Out of an Overcrowded Frying Pan into a Fire of Unsustainability?

Andrew MacLaran

Introduction

Cities accommodate a plethora of activities that relate to the productive, distributive, reproductive and exchange operations of our social system. Such functions possess different requirements and preferences with respect to urban location and building type. With their different degrees of market power, locational competition between these functions tends to create an urban landscape in which different activities become inscribed in geographical space. Thus, urban areas comprise a complex arrangement of different types of functional space. These specialised functional areas (industrial, commercial and residential) are linked with one another through transport and communication infrastructures, facilitating the movement goods, information and people, permitting the daily assembly of the labour force at its place of employment and its return home to recuperate.

Although over the long term, spatial competition tends to promote the physical separation of different types of function, the lengthy period required to amortise costly investments embodied in the built environment has meant that considerable spatial mixing of functions frequently remains apparent at a detailed geographical scale (see MacLaran, 2003). However, to a significant degree, the marked tendency towards the separation of activities has become reinforced, codified and even directly promoted by the prescriptions and operations of modern urban planning. As planners reacted to the heterogeneity of land uses which had arisen in the historic city and which often caused negative spill-over effects from the proximity of incompatible land uses, zoning policies commonly sought to create mono-functional urban districts.

In the American context, Scott (1980) has reviewed the long-term reorganisation of inner-city land uses and the redevelopment of central areas for more lucrative commercial functions during the twentieth century. He noted that this re-ordering and upgrading became possible only through the release of pressure on intra-urban locations made possible by the relocation of lower-value industrial operations and residential functions to suburbia, promoted by a policy of aggressive peripheral urban expansion facilitated through heavy investment in transport infrastructure. However, this led to widening of commuting fields and a wholly illogical arrangement of urban land uses where office workers commuted over increasingly lengthy distances to central city places of work while engendering reverse commuting of inner-city residents to suburban industry.

The process of urban development may be undertaken directly by the public sector, particularly of those elements too risky for the private sector to undertake or which lie beyond the logic of individual capitalists to provide. These include transport infrastructure or housing for poorer groups, and, indeed, the development of complete new towns. However, it is normally the private sector that is responsible for the greater part of urban real-estate development. It is a process which is therefore heavily influenced by powerful property interests, commercial companies and investment institutions. While planning systems seek to control, shape and influence the spatial outcomes of the development process, Feagin (1983, p. 3) has emphasised the key role of profitability in the creation of the urban environment and that urban space is not just a neutral container of societal functions:

> Cities under capitalism are structured and built to maximize the profits of real estate capitalists and industrial corporations, not necessarily to provide decent and liveable environments for all urban residents.

This contradiction between public good and private profit is particularly well exemplified by the creation of 'Edge Cities' which may provide short-term development profitability but which fail to meet social criteria of long-term sustainability.

The optimal functioning of an urban economy requires that land-uses are organised and circulation operates as efficiently as possible. The geography of employment, the residential location of different types of workforce and the existence of appropriate transportation infrastructures are therefore matters of considerable long-term importance for the sustainability of urban land-use patterns. Arrangements which minimise the necessity for intra-urban movements are more likely to be sustainable and may even generate a competitive edge over those urban economies obliged to invest in large-scale costly infrastructural transport programmes to accommodate the diurnal commuting necessary to address inappropriate land-use patterns.

This chapter reviews the development of suburban Dublin and, because of the increasing significance of office employment in the economy, thereafter concentrates on matters relating to the suburbanisation of office development in the city during the last decade of the twentieth century. Although it might be maintained that office suburbanisation presents the prospect of 'bringing office employment closer to its residentially suburban workforce', the chapter concludes that the creation of substantial amounts of office employment at the periphery and the effective development of an 'Edge City' is likely to generate enormous problems when viewed from a long-term perspective of urban sustainability.

Suburbanising Dublin

The second half of the twentieth century witnessed the profound transformation of Dublin from an essentially compact and densely-populated city totalling fewer than 700,000 inhabitants to a sprawling metropolis of more than a million residents. As late as 1961, the inner city had accommodated almost 160,000 of the 719,000 residents of the Dublin Region. By 1991 the inner city's residential population comprised only 76,500 out of a metropolitan total of 1,025,300.

Horner (1999) has identified several distinct types of city which have dominated Dublin's character through the twentieth century. Initially, 'slum city' was predominant during the early decades of the century. Its character was determined by the poor and densely-populated central area where a high proportion of the residents lived in overcrowded tenement dwellings. These had been carved out of the former dwellings of the city's élite and merchant classes who, during the nineteenth century, had increasingly abandoned the insanitary core in favour of more salubrious suburbs.

'Garden City', or more correctly the expanding ring of low-density suburban developments, came to dominate the character of Dublin by mid-century. It catered for the gathering pace of movement out of the core. On one hand, the demand emanated from private households seeking better housing and environmental conditions and, on the other, was associated with public-sector slum clearance programmes and the desire to reduce central Dublin's appallingly high residential densities.

In this process of suburbanisation, Dublin Corporation's role was highly significant. From a faltering start in the first decade of the century, marked by developments at Clontarf and Inchicore, suburbanisation proceeded apace during the 1920s, schemes of over 400 dwellings being developed in Marino, Drumcondra, Donnycarney and Cabra (see Figure 4.1).

During the following decade, the scale of development increased even further with the development of 3,200 cottages at Crumlin. However, it is interesting to note that even at mid-century Dublin remained a rather compact city. The private-sector development of the 'garden suburb' of Mount Merrion was developed from the 1930s to the 1950s, but was still geographically distinct and separated from the city itself despite its being just 6 km. from the city centre (MacLaran, 1993).

After World War II, while speculative private-sector residential development dominated across a broad swathe of newly created southern suburbs, from Rathfarnham, Churchtown, Dundrum through Stillorgan to Dean's Grange and Killiney, development elsewhere was much more mixed, associating low-income owner occupation and large elements of social housing. In the immediate post-war era, the Corporation's role was of major importance, marked especially by the

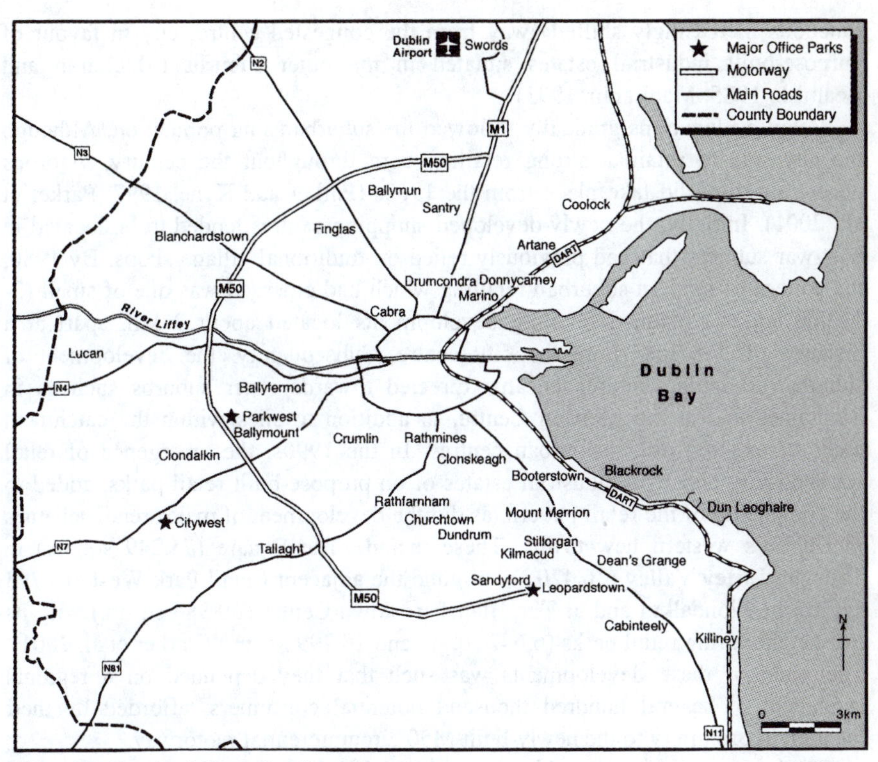

Figure 4.1 Dublin suburban locations

development of Artane and Ballyfermot in the 1950s. Indeed, by 1959, almost 45 per cent of the 42,360 dwellings built in Dublin since World War II had been developed by the Corporation (MacLaran, 1993).

In the following decade, public housing was developed over a wide area to the north of the city stretching from Finglas and Ballymun to Coolock. From the 1970s, the peripheral expansion of the city was confirmed by the policy to develop three 'new towns', expanding small pre-existing villages at Tallaght, Lucan and Clondalkin, and at Blanchardstown. The expansion of low-density suburban single-family residential areas conveniently maximised the demand for items of personal consumption, from household consumer durables to privatised transport. Capitalist productive industry was well served by the growing demand from potential consumers, particularly for motor cars, as life in suburban 'Arcadia' became unthinkable, and virtually impossible, without access to a car.

Meanwhile, this suburban residential growth became reinforced by the suburbanisation of other urban functions. From the 1960s, the demand for large single-storey industrial buildings and changes in transport technology, notably the increasing use of heavy trucks, in turn encouraged and facilitated a transformation of the city's industrial (manufacturing and warehousing) geography. Industrial functions increasingly shifted away from the congested central city in favour of purpose-built industrial estates situated in the outer suburbs (MacLaran and Beamish, 1985; MacLaran, 1993).

Retailing functions gradually followed the suburbanising population. Although the city was to retain a strong retailing core throughout the century, vigorous decentralisation did take place from the 1960s (Parker and Kyne, 1997; Parker et al., 2001). Initially, the newly-developed shopping centres tended to be located in post-war suburbs that had previously relied on traditional village shops. By 1980, the pattern of modern suburban retailing which had emerged was one of small (2-12,000 sq. m.) planned retailing developments located about 3 km. apart at a distance of 5–8 km. from the city centre. Subsequently, the development of suburban shopping centres became directed towards inner suburbs such as in Rathmines and at the Merrion Centre, in addition to sites within the catchment areas of existing outer suburban centres. In the 1990s, the emergence of retail warehousing, either on industrial estates or on purpose-built retail parks, added to the complexity of the retail pattern, as did the development of major retail schemes in Dublin's western new towns. These include The Square (75,249 sq. m.) in Tallaght, Liffey Valley (36,420 sq. m.) and the adjacent Retail Park West (15,793 sq. m.) in Clondalkin and at The Blanchardstown Centre (60,385 sq. m.) with its two neighbouring retail parks (6,642 sq. m. and 18,299 sq. m.) (Parker et al, 2001). The scale of these developments was such that they depended on a regional catchment of several hundred thousand potential consumers, afforded by their location in proximity to the newly-built M50 circumferential motorway.

To this complexity of residential, industrial and retailing decentralisation has been added, in more recent times, a significant degree of suburbanisation in office development, notably since 1995. It is to these trends and their ramifications that the chapter will now turn. It is an examination which will be undertaken in some detail, reflecting the growing domination of office-based employment in the urban economy and the importance for sustainability of the spatial relationship between the geography of different types of urban economic functions and the location of appropriate labour forces. A concern for the long-term sustainability, or energy efficiency, of cities requires an examination of the contemporary macro-organisation of urban land use because the intense separation of urban functional spaces which is becoming increasingly evident is sustainable only with the expenditure of immense quantities of energy.

Office Suburbanisation

With their generally high degree of accessibility for workforce and clients and with the benefits for personal and business interaction afforded by the clustering of operations, city centres had long dominated the location of offices. Dublin was unexceptional in this regard. Until the 1960s, most of the city's office functions were accommodated either in buildings adapted from residential functions, as in

Dublin's eighteenth-century streets and squares, or in small-scale purpose-built offices developed during the late nineteenth century. However, irrespective of building type, office locations were overwhelmingly concentrated in the city centre.

The economic expansion of the 1960s generated a growth in office employment and a demand for additional accommodation. This continued to be furnished in significant by the continued conversion of much of the eighteenth-century housing stock situated in prestigious areas of the inner city, around St. Stephen's Green, Fitzwilliam and Merrion squares. However, new office development schemes started to make their appearance from the early 1960s. These tended also to be located largely in the city's central area, notably in the vicinity of Trinity College at College Green and Hawkins street, and also around St. Stephen's Green. A few developments did venture tentatively into the inner suburb of Ballsbridge but, by the end of the 1960s, only 3 of the 41 completed post-1960 office developments were located outside the city's established office core of Dublin 2 and its immediate fringe comprising Dublin 1 and Dublin 4. Figure 4.2 shows the location of Dublin's postal districts, which provides the spatial basis for analysing development.

During the 1970s, 140 office developments reached completion in Dublin. While their geographical spread continued to widen, fewer than 10 per cent of these were situated outside the Dublin 2 office core or its immediate fringe in Dublin 1, 4, 7 and 8. Indeed, almost 70 per cent were located in Dublin 2 itself.

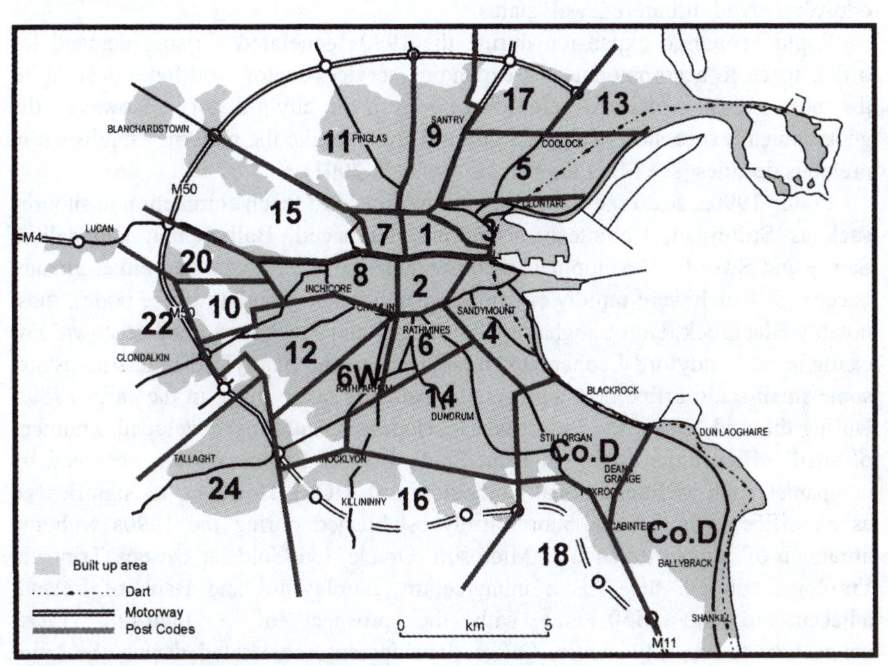

Figure 4.2 Dublin postal districts

By the end of the 1970s, although the city's outer southern suburbs of Stillorgan, Dundrum, Cabinteely, Dun Laoghaire, Leopardstown, Booterstown and Clonskeagh could each record the presence of one or possibly two modern office developments, no suburban proto-nodes had yet begun to emerge and only one modern office building was located in the northern outer suburbs.

In the 1980s, a further 382,100 sq. m. of office space was developed, with some 46,840 sq. m. (12.3 per cent of the total) being located in the outer suburbs. During the early part of the decade, as the second office boom reached its height, Dun Laoghaire and Blackrock emerged as focal points for development. Yet, by the end of the 1980s, when the city-wide stock of modern office space totalled 975,285 sq. m. in 365 developments, some 75 per cent of this space was located in the prime office locations of Dublin 2 and the inner suburb of Dublin 4. A further 16 per cent was located in the inner-city office fringe in Dublin 1, 7 and 8. The remaining suburbs accounted for just 9 per cent of the stock.

Legislation in 1986 created fiscal incentives for property-based renewal in certain designated areas of the city. Initially, these reinforced developers' preference for inner-city sites, although within the central area they did encourage a widening of the geography of office development into parts of Dublin 1, 7 and 8 which fringed the office core, notably at the Custom House Docks, along the Liffey's quays, around Christchurch and between Aungier street and Golden Lane. By the 1990s, these incentives had also induced a limited amount of office development in the new town of Tallaght where a greenfield site had, somewhat oddly, received 'urban renewal' status.

Rapid economic expansion during the 1990s generated a rising demand for office space to accommodate the expanding services-sector workforce, leading to the most intensive office development boom in the city's history . However, the geographical expression of this activity was quite unlike the patterns established in previous decades (see MacLaran and O'Connell, 2001).

In the 1990s, scattered developments continued to reach completion in suburbs such as Stillorgan, Cabinteely, Dundrum, Kilmacud, Ballymount, Clondalkin, Santry and Swords. Development also continued apace in what had either already become or which were rapidly emerging as significant suburban office nodes, most notably Blackrock, Dun Laoghaire, Clonskeagh and Sandyford-Leopardstown. For example, in Sandyford-Leopardstown, adjacent to the Sandyford industrial estate, some small-scale office developments had already taken place in the early 1980s. During the mid-1980s, the Industrial Development Authority developed a number of small office units at the adjacent South County Business Park, occupied by companies such as International Computers Ltd. (ICL). However, its significance as an office node became more firmly established during the 1990s with the attraction of companies such as Microsoft, Oracle, LG Goldstar Design, Trintech, Eurologic, Eircell, the AIB training centre, Barclaycard and Bank of Ireland, adjacent to the M50 and with the prospect of a light-rail (Luas) connection to the city centre. Office development accelerated during the latter half of the decade and, at the height of the fourth office development boom, increasing numbers of industrial units were demolished to make way for office blocks while other development took place on greenfield sites. By December 2003, a total of 167,336 sq. m. had been developed. Many of these new blocks were of substantial size, the Atrium Buildings I and II each exceeding 13,260 sq. m. and being amongst the largest stand-alone office blocks in Dublin (Figure 4.3). To the south of the Leopardstown Road, Treasury Holdings' development of its Central Park office complex will, on completion, accommodate over 160,000 sq. m. of space. This represents almost 50 buildings the size of Liberty Hall in central Dublin (Figure 4.4).

New proto-nodes also appeared in the inner-suburb of Dublin 3, with the development of the East Point Business Park, and in the outer suburbs at Tallaght, Citywest, along the Nangor Road and at Blanchardstown and Loughlinstown. East Point, Tallaght and Parkwest (Figure 4.5) availed of fiscal incentives to encourage development. Such was the scale and rapidity of development activity in these new locations that by the end of 2000, East Point comprised Dublin's largest suburban office node, surpassing more established suburban centres at Blackrock, Dun Laoghaire and Clonskeagh.

During the period 1990-2000, less than 20 per cent of the office space developed in the city was situated in Dublin 2, with a further 8 per cent being located in Dublin 4. In contrast, the outer suburbs accounted for 53 per cent of the total. As a result of this unprecedented scale of the development and its changed geographical focus, the distribution of the city's modern office stock was transformed. During 2001, the peak year for office completions, only 28 per cent was located in the central area, comprising the office core in Dublin 2 and its inner-fringe located in Dublin 1, 7 and 8 and the inner suburb of Dublin 4.

Figure 4.3 Atrium I and II, Sandyford, Dublin 18

Figure 4.4 Liberty Hall, Dublin 1 (1965)

Figure 4.5 Park West Business Park, Dublin 22

This altered development geography had a profound impact upon the location of the modern office stock and, therefore, upon the geography of office employment in Dublin. As late as 1995, over 85 per cent of the office stock was still located in the traditional office core in Dublin 2 (55 per cent), its inner overspill areas of Dublin 1, 7 and 8 (17 per cent) and the prestigious inner suburb of Dublin 4 (13 per cent). Table 4.1 shows that by the end of 2003 little over 35 per cent of the modern stock was located in Dublin 2, with Dublin 4 accounting for a further 11 per cent. However, the outer suburbs, with a stock of over 637,000 sq. m., accounted for over 36 per cent of the total, matching the 1980 figure for the total city-wide stock of modern office space.

Furthermore, with respect to the location of office-type employment in the metropolis, these calculations do not take account of the estimated additional 105,000 sq. m. of modern office space which has been developed in association with industrial management operations, tele-sales and support service functions. This stock is located overwhelmingly on industrial estates at peripheral locations (MacLaran, 1999; MacLaran, 2001).

The factors underlying this large-scale suburbanisation of office development are complex and only the more significant elements can be noted here briefly. These lie in the changing planning policy environment, administrative fragmentation of the metropolis, changing user demand factors and evolving preferences of developers and investors (see MacLaran and Killen, 2002; Bertz, 2002a, 2002b). Disturbingly, it seems that public-sector policies have played a major role.

Greater levels of protection for historic buildings and planning restrictions that hindered the construction of large-scale developments in the city centre, together with a growing degree of competition for available sites from alternative functions such as hotels and apartment developments, resulted in a growing paucity of suitable office development sites in the city centre. In contrast, development at the periphery provided ease of site assembly and was encouraged by relatively liberal planning regimes which reflected the desire of suburban local authorities for additional income from the commercial rates (property-based taxes) which such office developments would convey (Bertz, 2002a, 2002b).

Table 4.1 Location of the modern office stock, December 2003

Area	*Stock (sq. m.)*	*Percentage*
Dublin 2	831340	35.20
Dublin 4	268915	11.38
Dublin 1,7,8	248037	10.50
IFSC	148095	6.27
Blackrock/Dun Laoghaire	92003	3.90
North suburbs	161197	6.82
West suburbs	253871	10.75
Other South suburbs	358621	15.18
Total stock	**2362079**	**100.0**

Source: HOK/CURS TCD office database, 1960-2003

Simultaneously, there also emerged a new type of demand for office space, generally derived from an international client with few locational constraints but demanding large (>10,000 sq. m.) amounts of space, which could not be developed in central Dublin. More generally, at a time when staff recruitment and retention issues were becoming increasingly important to companies during a period of full employment in the economy, access by its overwhelmingly suburban workforce became a growing consideration in the selection of accommodation (Bertz, 2002a, 2002b). Suburbia could offer lower unit costs and provide higher car parking to floorspace ratios in attractive office complexes or science park environments. The availability of tax incentives for property development (see above) in certain suburban locations was yet another inducement.

Finally, from an investment perspective, at a time of rapidly rising values, investors became increasingly keen to secure additions to their office property portfolios. As the scale of development in the city centre had been reduced to a historically low level of activity during much of the 1990s, investors became obliged increasingly to consider more peripheral developments in order to satisfy their investment demand.

The overall impact of this interpenetration of a changed environment with respect to urban planning policy, the growing demand for space by investors and altered preferences of users, together with the relative ease with which suburban office development could be effected, changed profoundly the geography of office employment in Dublin. Indeed, within a matter of two decades, a new geography of highly dispersed employment has been shaped. Set amid low-density suburban residential areas, this new Edge-City comprises a landscape of retail centres, medium-rise high-specification office blocks and light-industrial units.

'Edge City' and Sustainability Considerations

Movements between spatially separate functional areas in cities account for a significant level of energy consumption. The optimal functioning of the urban economy therefore requires that land-uses are organised and circulation operates as efficiently as possible. Thus, the configuration of urban land uses, notably the location of different types of residential area and types of workforce, the geography of employment and the existence of appropriate transportation infrastructures constitute matters of key concern to urban analysts and are of considerable importance in evaluating the long-term sustainability of urban land-use arrangements.

By the late 1990s, Dublin had been transformed into a sprawling metropolis with a population of over 1 million persons, the majority of whom were suburban dwellers. It had become, at least in skeletal form, a multi-nodal metropolis marked by the emergence of an 'Edge City' (Horner, 1999; Williams and Shiels, 2000, 2002) which comprised a series of industrial and commercial nodes of employment, strung like beads along the necklace of the M50 motorway, tied together in a Gordian knot of reliance upon a complex pattern of car-based commuting. It extended from Loughlinstown, Sandyford and Leopardstown in the south east, though the western spread of development focusing around the new towns of Tallaght (City West and Baldonnell), Clondalkin (Newlands Cross, Nangor road, Palmerstown and Quarryvale) and Blanchardstown (Ballycoolin and Cappagh Roads), to the northern fringe around the airport (Santry-Swords) and the nascent developments at Balgriffin and north fringe.

Gradually, however, the generally market-led trends in land-use and development which have dominated Dublin's development and which facilitated the decentralisation of employment to suburbia, have given rise to major problems. Not only have the peripherally situated workplaces developed largely in the absence of adequate infrastructures, but there has been little consideration of how these can *ever* be serviced adequately. Public transport systems work best where they connect a multiplicity of residential origins with a single destination, normally the city centre. However, the evolving pattern of primarily market-led development in Dublin has created a multiplicity of employment nodes and potential destinations. The consequence has been a growth of inter-suburban commuting and increasingly complex pattern of journeys to work which cannot be readily accommodated by public transport systems. The inadequacy of Dublin's public transport services to the emerging employment nodes of 'Edge City' has been documented by MacLaran and Killen (2002). The pattern of development has created a high level of enforced dependency on the motor car for commuting. This, in turn, has generated suburban traffic congestion, which inevitably encourages demands for expensive road improvements and motorway construction to facilitate movement within and between suburbs.

In the creation of this intractable problem, a major factor has been the planning, administrative and urban financing contexts within which development has taken place. In order to save cash in the short term, the central government failed during the 1980s and 1990s to maintain the real value of its subvention to local authorities after the abolition of domestic rates (locally-based residential property taxes) in the 1970s. This, in turn, often induced impecunious local authorities to facilitate and attract office developments to peripheral locations in order to increase their income from commercial rates.

Increasingly, land which once accommodated single-storied industrial buildings is being redeveloped for multi-storied office buildings, vastly increasing the quantity of floorspace on individual sites. However, this has created major difficulties with respect to commuter access. Industrial buildings generally have a far larger quantity of floorspace per employee than do offices. Sites which may have accommodated a workforce of a few dozen industrial operatives or half a dozen warehousemen may subsequently accommodate more than a hundred office workers. Severe traffic congestion has been the inevitable impact of the rapid increases in the number of commuters travelling to such suburban sites. Thus, ironically, the taxpayer now faces the prospect of having to pay for costly suburban road improvements in order to alleviate congestion created by commuters driving to the newly-developed suburban places of employment which probably should never have been created in these locations at all.

In proposing their schematic model for the organisation of a public transport system to cater for the suburbanisation of office employment, MacLaran and Killen (2002) have concluded that the number of suburban nodes must be strictly limited and that such office developments should be located in the vicinity of retail complexes to provide additional justification for more than a skeletal public transport service throughout the day. Only then will the macro-scale land-use arrangements which have emerged at Dublin's periphery become more 'sustainable'. Failure to adopt such a policy will render it virtually impossible for public transit to cater for the multi-origin multi-destination patterns of inter-suburban travel which are emerging.

Dublin is, of course, far from unique in its trend towards the peripheralisation of economic activities and employment. Los Angeles is the archetypal poly-nucleated metropolis which underwent such an intensity of peripheral development around its scattered component towns and villages that a traditional commercial core was barely recognisable. Yet even a city such as Paris, which possesses a strongly iconic core charged with historical and cultural significance, had by the late twentieth century evolved into a sprawling metropolis in which a minority of employees worked in the centre. Indeed, trends in Paris provide a salutary warning with respect to the patterns of vehicular movement which are likely to emerge as a result of vigorous peripheral expansion. Excluding pedestrian movements or those by bicycle, vehicle-base journeys between Parisian suburbs rose from 58 per cent of the total in 1978 to 70 per cent in 1998. This is highly significant because, as Table 4.2 reveals, a far lower proportion of inter-suburban trips are undertaken by public transport and that fewer than 10 per cent of vehicular-based trips which both originated and terminated in the outer suburbs were effected by public transport.

Yet, even as Dublin's 'Edge City' was emerging, the creation of sizeable nodes of employment at the city's periphery resulted in a widening of its

Table 4.2 Daily vehicle-based movement, Paris 1998

Origins/destinations	*Percentage of all trips*	*Percentage by public transport*
Outer suburb-Outer suburb	37.2	9.8
Outer suburb-Inner suburbs	8.6	22.5
Inner suburb-Inner suburb	24.6	21.4
Outer suburb-City centre	5.1	65.0
Inner suburb-City centre	11.3	56.2
City centre-City centre	13.2	66.8

Source: *Atlas des Franciliens 1. Territoire et Population*, INSEE (2000), Paris.

'footprint' as long-distance commuting, often by car, became common. 'Edge City' was far more accessible to beyond-the-edge townships than had ever been the case for the city centre, except to those surrounding towns linked to central Dublin by rail. Indeed, 'Edge City' was often more accessible from distant towns, such as Carlow, Portlaoise or Navan, than to car-based commuters trying to effect a daily journey on heavily congested roads from Dublin's inner suburbs. By the late 1990s, Dublin's outer commuter belt extended for 90 km. from the city (Williams and Shiels, 2000, 2002) and towns, such as Arklow, Carlow, Portlaoise, Portarlington, Mullingar, Navan and Dundalk began to attract residential developments targeted at commuters to Dublin who sought an escape from the escalating property prices and declining affordability of dwellings in the city.

Significantly, for the daily assembly of their workforces, these newly-created employment locations rely heavily on commuting by private car. However, periodic crises in the distribution and pricing of fuel supplies underline the degree to which the modern metropolis has become dependent upon the efficient operation of its circulation space in a space-economy characterised by intense separation of land uses. As the prospect of impending suburban gridlock looms, the long-term viability of dispersed 'Edge-City' developments will be further challenged by the inevitable increases in the real cost of private transportation as the global availability of fossil fuels diminishes.

Notes

The data upon which this chapter is based has been drawn from the database of modern (post-1960) office developments in Dublin held by Hamilton Osborne King, Dublin, and The Centre for Urban and Regional Studies, Trinity College Dublin. It also draws on the Office Reviews compiled by those organisations annually since 1989.

References:

Bertz, S. (2002a) 'The growth in office take-up in Dublin's suburbs: a product of occupiers' changing locational criteria?', *Journal of Irish Urban Studies*, Vol. 1(2), pp. 55-75.

Bertz, S. (2002b) 'The peripheralisation of office development in the Dublin Metropolitan Area: the interrelationship between planning and development interests', *Irish Geography,* Vol. 35(2), pp. 197-212.

Feagin, J. R. (1983) *The Urban Real Estate Game: Playing Monolpoly With Real Money,* Englewood Cliffs: Prentice-Hall.

Horner, A. (1999) 'Population dispersion and development in a changing city-region', in Killen, J. and MacLaran, A. (eds) *Dublin and its region: Contemporary issues and challenges for the twenty-first century,*

Geographical Society of Ireland Special Publication, 11, GSI and The Centre for Urban & Regional Studies, Trinity College Dublin, pp. 55-68.

INSEE Institut National de la Statistique et des Études Économiques (2000) *Atlas des Franciliens 1. Territoire et Population*, Paris: INSEE Institut National de la Statistique et des Études Économiques.

MacLaran, A. (1993) *Dublin: the shaping of a capital,* London: Belhaven / Wiley.

MacLaran, A. (1999) 'Inner Dublin: Change and Development', in Killen, J.K. & MacLaran, A. (eds) *Dublin and its region: contemporary issues and challenges for the twenty-first century*, Dublin: Geographical Society of Ireland & CURS, Trinity College Dublin, pp. 21-33.

MacLaran, A. (2003) *Making Space,* London: Arnold.

MacLaran, A. and Beamish, C. (1985) 'Industrial property development in Dublin, 1960–1982', *Irish Geography*, Vol. 18, pp. 37-50.

MacLaran, A. and Killen, J. (2002) 'The suburbanisation of office development in Dublin and its transport implications', *Journal of Irish Urban Studies*, Vol. 1(1), pp. 21-35.

MacLaran, A. and O'Connell, R. (2001) 'The changing geography of office development in Dublin', in Drudy, P. J. and MacLaran, A. (eds) *Dublin: Economic and Social Trends - Volume 3*, The Centre for Urban & Regional Studies, Trinity College Dublin, pp. 25-37.

Parker, A. J. & Kyne D. M. (1997) *Dublin shopping centres: a statistical digest III,* Dublin: Centre for Retail Studies, UCD.

Parker, A. J., Kelly, F.M. and Kyne, D. M. (2001) *The Dublin Shopping Centre and Retail Park Digest*, Dublin: The Centre for Retail Studies, UCD.

Scott, A. J. (1980) *The Urban Land Nexus and the State,* London: Pion.

Williams, B. & Shiels, P. (2000) 'Acceleration into Sprawl: causes and potential policy responses', *Quarterly Economic Commentary,* June 2000, Dublin: Economic and Social Research Institute.

Williams, B. and Shiels, P. (2002) 'The expansion of Dublin and the policy implications of dispersal', *Journal of Irish Urban Studies*, Vol. 1(1), pp. 1-19.

Chapter 5

Urban Form and Reducing the Demand for Car Travel: Towards an Integrated Policy Agenda for the Belfast Metropolitan Area?

Malachy McEldowney, Tim Ryley, Mark Scott and Austin Smyth

Introduction

Much of the interest in promoting sustainable development in planning for the city-region focuses on the apparently inexorable rise in the demand for car travel, and the contribution that certain urban forms and land-use relationships can make to reducing energy consumption and emissions harmful to both local environmental and global ecological conditions. As Banister (1999) records, although some of the growth in car dependency can be attributed to the acquisition of cars by individuals, it also reflects the distribution of functional opportunities which have become more dispersed throughout the city-region. Within this context, policy prescription has increasingly favoured a compact city approach with increasing urban residential densities and mixed-use development to address the physical separation of daily activities and the resultant dependency on the private car. This chapter aims to outline and evaluate recent such efforts to integrate land-use and transport policy in the Belfast Metropolitan Area in Northern Ireland, wherein significant policy initiatives have recently been undertaken at both regional and metropolitan scales.

Although considerable progress has been made in integrating land-use and transportation policy for the Belfast city-region – both institutionally and conceptually – this chapter draws on two major recent research projects to argue that significant challenges remain to be overcome. In particular, the results of an Engineering and Physical Sciences Research Council Sustainable Cities project (EPSRC)[1] underlines the extent of existing car dependency in the metropolitan area and prevailing negative attitudes to public transport. Furthermore, drawing on the authors' involvement as facilitators for the public consultation process for the Belfast Metropolitan Area Plan (BMAP),[2] this chapter suggests that although sustainability principles are unquestioned at the city-wide scale, at the level of implementation sustainable development becomes a contested concept. In particular, the chapter explores the widespread neighbourhood and community opposition (NIMBY-ism) to increasing housing densities and brownfield development during the formulation of BMAP. Accordingly, this chapter is structured as follows: firstly, the chapter will locate the discussion within the context of sustainable development and urban form. Secondly, a brief overview of land-use and transport trends in the case study area will be provided, followed by an assessment of recent attempts to integrate land-use planning and transport policy, at both regional and metropolitan scales. The remainder of the chapter will focus on emerging challenges to implementing a sustainable urban development agenda, in particular examining the micro-politics of the sustainable city debate before conclusions are developed relevant to the future practice of land-use transport policy integration.

Sustainable Development and Urban Form

The phenomenon of tension between a compact city structure and the motorcar is not a recent one, but dates to the 1970s and before (see Owen, 1972; Hass-Klau, 1990). Much of the urban environment has been developed to accommodate the motorcar, including the road network, filling stations and car parks, and the growth of private motorcar ownership has enabled city residents to migrate to the suburbs and beyond. Indeed, virtually all Western countries have witnessed deep-seated counterurbanisation trends in the post-war period (Breheny, 1992). As noted by Gaffikin and Morrissey (1999, p.96):

> …these processes were the outcome of 'push' factors from central cities, such as congestion, pollution, noise, crime and grime, together with 'pull' factors in the suburbs, such as the facility of under-priced car use, good supply of relatively cheap land, public policies which subsidized housing and highways, technology changes from the telephone onwards which minimized the effect of distance, and the desire by employers for low-rise offices which allowed for more efficient employee interaction.

In many advanced capitalist societies in recent times, land-use trends alongside the opportunities afforded by past road construction efforts and the mass availability of private cars has led to the breaking of traditional relationships between home, work and leisure opportunities (Vigar, 2002). Urban sprawl has resulted from increasingly affluent householders and commercial investors exercising their locational choices in a free market, aided by the availability of good quality transport infrastructure and relatively cheap private transport. As identified by Banister (1998) transport has been seen as a principle permissive factor in the development of suburban housing and in redirecting development pressures away from the city centre. This dynamic is reinforced by the development of out-of-town and edge-of-town growth in retail, business and leisure services at motorway intersections and along bypasses. However, recently in the UK and Ireland, following trends in mainland Europe, the notion of a dense and compact city with mixed-use neighbourhoods has gained a high political currency. The Commission of the European Communities' Green Paper on the Urban Environment (1990, p. 60) strongly advocates a 'compact city' solution to mounting environmental problems, proposing that "strategies which emphasise mixed use and denser development are more likely to result in people living close to work places and the services they require from everyday life". In the UK this approach has been formalised in the Urban Renaissance Report (Urban Task Force, 1999) and subsequent White Paper and in the Republic of Ireland these sentiments have been translated into formal guidance through Planning Guidelines on Residential Density (DoELG, 1999) and the National Spatial Strategy (DoELG, 2002), which emphasise the need to consolidate urban areas and to concentrate residential development in locations where it is possible to integrate employment, community services, retailing and public transport.

A central theme in recent policy prescriptions has been the integration of land-use planning and transportation. There is a growing realisation that the primary means to improve both the environment and congestion is to reduce the need to travel, and in particular the length of trips needed to carry out daily activities. In this context the planning system has a key role. As Banister (1997, p. 447) suggests, "to reduce levels of car dependence and trip lengths, planning decisions must have an instrumental role through establishing and implementing clear development principles based on sustainability". The whole debate concerning urban form, transport, and quality of life coincides with a longer standing debate in planning about the *accommodation of development* (Breheny, 1992). Indeed, this discourse encapsulates three of the most important and interrelated issues facing land-use planning at the beginning of the twenty-first century: how to accommodate substantial growth in the number of households; how to revitalise cities; and how to create more sustainable urban areas (Heath, 2001). The compact city model, therefore, is supported for a number of reasons which relate to sustainability and include: conservation of the countryside; less need to travel by car, thus reduced fuel emissions; support for public transport and walking and cycling; improved access to services and facilities; more efficient utility and infrastructure provision; and, revitalisation and regeneration of inner urban areas (Burton, 2003).

To address these issues, plans and planners are increasingly moving beyond 'land-use' planning to embrace a wider agenda, adopting in effect the European notion of spatial planning instead of the traditional UK and Irish regulatory approach (Tewdwr-Jones and Williams, 2001). This new thinking seeks to achieve an integration of the social, economic and environmental dimensions, and to broker connections between the core city and its wider metropolitan hinterland. In Northern Ireland, this broader agenda resulted in the preparation of a new regional strategic framework for spatial development (*Shaping Our Future*), developed from a collaborative planning process involving widespread participation of community interests and multiple stakeholders (McEldowney and

Sterrett, 2001). This broader agenda notwithstanding, much of the debate inevitably focused on the distribution of forecasted housing growth and included tensions relating to brownfield versus greenfield development (especially dealing with intense housing pressure on Belfast's greenbelt), edge of settlement development versus new free standing 'village' concepts and the split in housing demand allocation between the Belfast region and the rest of Northern Ireland (Neill and Gordon, 2001). The remainder of this chapter will examine the outcome of this planning process, focusing on proposals for the Belfast Metropolitan Area as they evolved during the Area Plan process.

The Belfast Metropolitan Area

The Belfast Metropolitan Area (BMA) covers the administrative districts of Belfast City, Castlereagh Borough, Carrickfergus Borough, Lisburn Borough, Newtownabbey Borough and North Down Borough, and is the largest urban centre in the region with an estimated population of 650,000 people (DoE, 2001). Belfast has been described as one of the UK's most car dependant cities (Cooper et al., 2001). The last 25 years have seen huge increases in car ownership in the UK and while Belfast and Northern Ireland have lagged behind Britain in the level of car ownership, it is rapidly catching up. Between 1990 and 2000, for example, the number of vehicles licensed in Northern Ireland has increased by almost 35 per cent, while the comparable figure for Britain is just 15 per cent (DoE, 2001). Growth in car ownership has increased by around 11 per cent between 1995 and 2000, and by over 400 per cent since 1960, with the number of vehicles expected to double by 2025 (DRD, 2001). In terms of travel culture, the car is by far the most dominant mode for personal travel in the BMA – recent Department of Regional Development (DRD) (2002a) figures suggest that 81 per cent of the workforce travel by car, van or minibus in their journey to work. Modes that have less adverse impact on the environment, such as walking, cycling and public transport, have a combined share of only 19 per cent. The result of this car dependency has inevitably been large increases in road traffic. In the BMA, peak hour traffic conditions are steadily worsening, causing accessibility and environmental problems – for example, between 1995 and 2000 the Department of Environment highlight that increases in traffic flow measured on some of the key routes to and from Belfast City Centre has risen by up to 20 per cent (DoE 2001). The use of the car for many journeys is encouraged by the availability of free or low cost car parking at key destinations. Trends in the availability of public car parking spaces in Belfast City Centre over the last 20 years show almost a five-fold increase in the number of spaces available (DoE, 2001). In Northern Ireland as a whole, people who have cars or access to cars are making more and longer journeys. Total distance travelled in the Region by car has increased from an estimated 4.5 billion km in 1971 to over 15 billion km by 1998 (DRD, 2002b).

Reinforcing these trends has been the emergence of a low-density urban form characterised by population loss in the central city. The BMA has experienced a massive out-migration from the urban core, common to most UK cities, although exacerbated by the 'Troubles' (Cooper et al., 2001). There was a period of intense urban decentralisation of Belfast between 1971 and 1981 when the population fell from 416,700 to 314,300 (Ellis and McKay, 2000). It continued to decline throughout the 1980s and by 1991 it had fallen to 279,200, as people continued to move out of the city into the contiguous suburbs and dormitory towns and villages. Ellis and McKay (2000) suggest that although a significant degree of population movement occurred as a result of the violent political conflict, there were a number of other reasons for this population shift including: the redevelopment of the inner city; the decline in mean household size; rising car ownership resulting in increased commuting; regional policy, which encouraged out migration from Belfast. However, since 1991, there has been a small growth in the population within the Belfast City Council area following decades of decline (DoE, 2001) in addition to the continued growth in the outer districts.

Current Policy Context: Towards Sustainable Development?

In recent years, land-use and transportation planning have enjoyed a high profile within Northern Ireland, underlining the current quest to secure more effective linkages between these two areas of public policy. The current regional planning framework in Northern Ireland is provided by *Shaping Our Future: The Regional Development Strategy* (RDS) 2025 (DRD, 2001), a statutory plan prepared by the Department of Regional Development (NI) and endorsed by the Northern Ireland Assembly in 2001. The broad aim of the spatial strategy is to guide future development in order to "promote a balanced and equitable pattern of sustainable development across the Region" (p. 41) and a framework of interconnected hubs, corridors and gateways is adopted. As Chapter 1 indicated, two regional *gateways* are identified – Belfast and Londonderry/Derry – in addition to a polycentric network of *hubs* based on the main regional towns serving a strategic role as centres of employment and services for urban and rural communities. The key and link transport *corridors* provide the skeletal framework for future physical development (Figure 5.1).

In relation to the Belfast Metropolitan Area, the RDS outlines the main thrust of its approach as a balance between concentration and decentralisation. Thus, the development of the Belfast city-region is based on a strategy of encouraging the revitalisation of the BMA; providing for major planned lateral expansion of the key transport corridors to the south and north of the city; developing the surrounding District Towns as counter magnets to the Metropolitan Area; and the accommodation of 'overspill' growth from the BMA by the expansion of seven nearby smaller towns (Figure 5.2). Perhaps the key challenge outlined by the RDS will be the accommodation of the projected housing growth for the BMA.

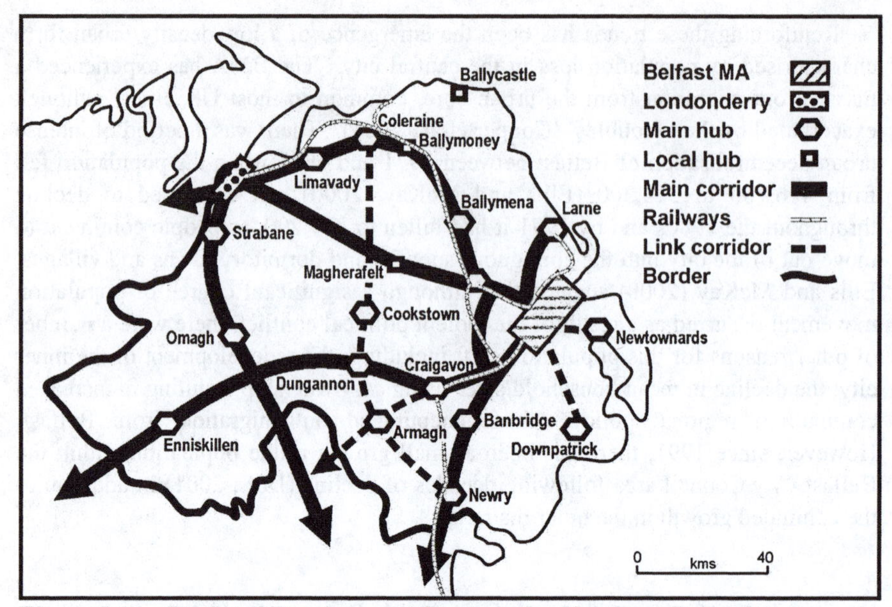

Figure 5.1 Regional Development Strategy for Northern Ireland

Out of a regional need of 160,000 dwellings for this period, the Strategy has allocated 51,000 to the Districts covered by the BMA, of which 42,000 should be located within the existing urban footprint. The RDS, therefore, at least in rhetoric, supports the concept of the 'compact city' as a means of reducing the need to travel, establishing a regional target of 60 per cent of new housing to be located within existing urban areas (which contrasts with the recent level of achievement of less than 30 per cent). It suggests a focus on brownfield sites within the city and marks a significant policy shift from previous patterns of greenfield and edge of city residential development. This is consolidated further in the Strategy's transport section which places emphasis on the creation of a Metropolitan Transport Corridor Network radiating out from the city centre and characterised by improved public transit services and increased density residential development along the corridors. The physical separation of key land-uses is also addressed with policies aimed at promoting mixed-use development. These themes have been further developed in the formulation process of the Belfast Metropolitan Area Plan (BMAP), currently under preparation by the Department of Environment (NI). However, although the RDS attempted to move spatial planning beyond land-use planning concerns, it is likely that the key issue in the formulation of BMAP will be the accommodation of future housing growth. Certainly, this issue has been central in the public consultation process (Gaffikin et al., 2003) discussed in detail below.

Parallel to the publication of a regional spatial framework, the Department of Regional Development also prepared a Regional Transportation Strategy (RTS)

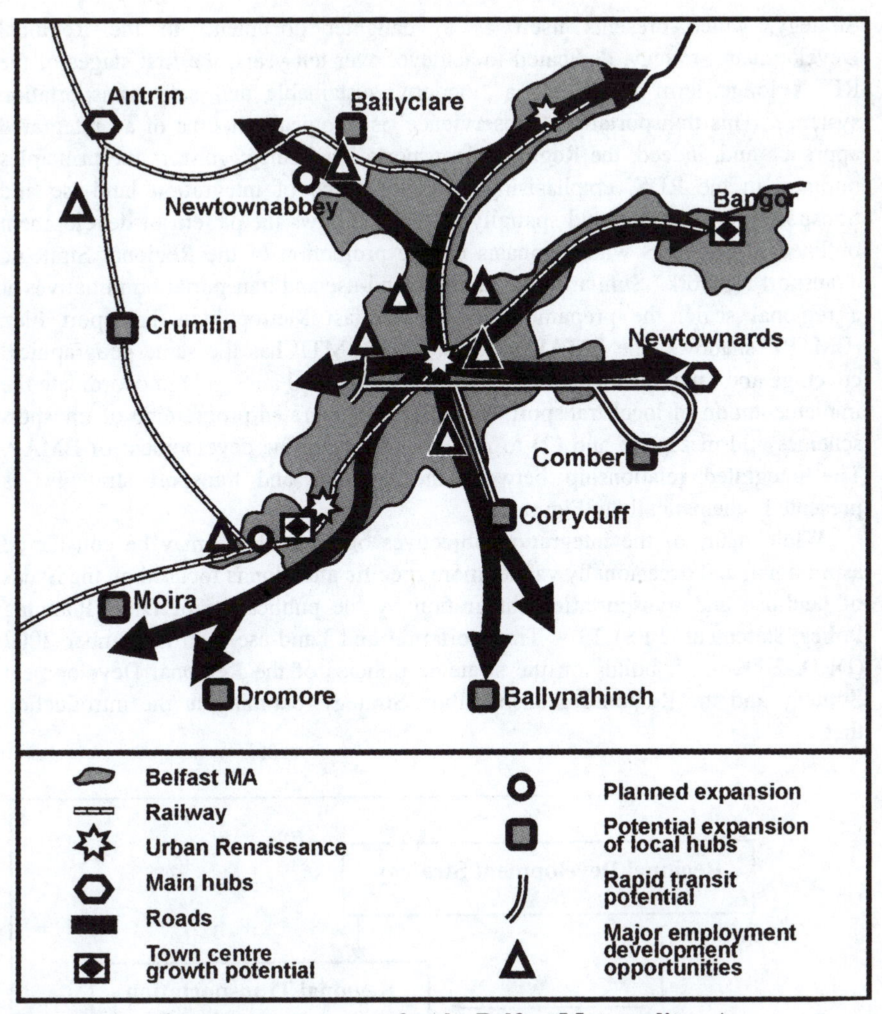

Figure 5.2 Development context for the Belfast Metropolitan Area

(DRD, 2002), approved by the Northern Ireland Assembly in July 2002. The aim of the RTS is to support the RDS and to move over a 10-year period towards the achievement of a longer-term transportation vision, which promotes a more sustainable transport modal use. At a strategic level, the integration of land-use and transportation responsibilities within the Department of Regional Development is a positive acknowledgement of the inter-dependant relationship between these two functions. Effective integration demands clarity in the hierarchy of power, with planners arguing the case for planning's overarching responsibility, within which transportation should play a key, but subservient role. This, surprisingly, is accepted and promoted in the Regional Transportation

Strategy, which presents itself as a 'daughter document' to the Regional Development Strategy, dedicated to achieve, over ten years, the first stages of the RDS's longer-term vision for a 'modern, sustainable and safe transportation system'. This transportation 'subservience' is a logical outcome of an integrated approach and, indeed, the Regional Transportation Strategy mirrors the principles outlined in the RDS, emphasising the importance of integration land-use and transportation planning, and spatially the RTS follows the pattern of development outlined in the RDS with emphasis on the promotion of the Regional Strategic Transport Network. Similar to the parallel land-use and transportation initiatives at a regional scale, the preparation of the Belfast Metropolitan Transport Plan (BMTP) 'shadows' the BMAP process. The BMTP has the same geographical coverage and timeframe as BMAP, and has two general aims: (1) to coordinate the implementation of local transport, outlining an integrated programme of transport schemes and measures; and (2) to reflect and inform the development of BMAP. The integrated relationship between the land-use and transport strategies is presented schematically in Figure 5.3.

While many of the integration objectives outlined above may be considered aspirational and occasionally vague, more specific attention is focused on the issues of land-use and transportation integration by the publication of Draft Planning Policy Statement (PPS) 13 – Transportation and Land-use – in December 2002 (DRD, 2002b). It builds on the strategic policies of the Regional Development Strategy and the Regional Transportation Strategy, claiming in the introduction that:

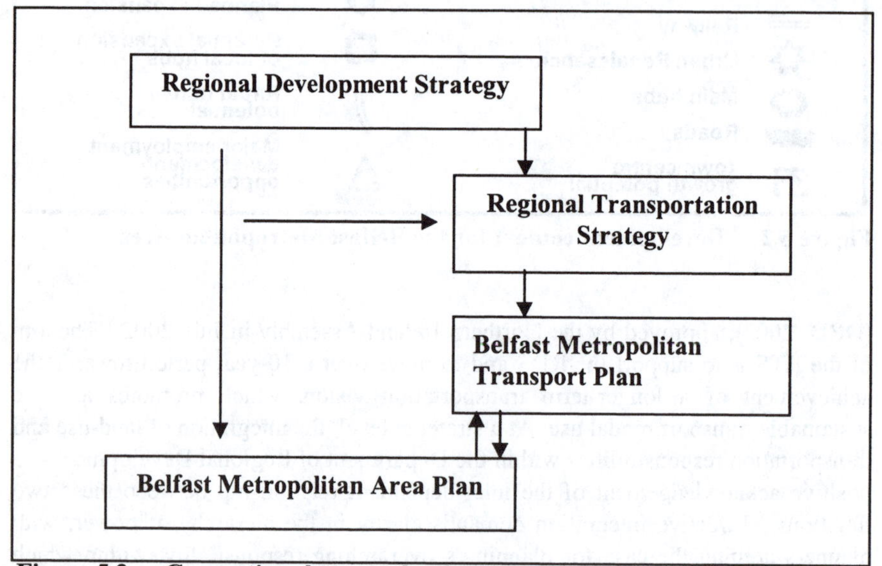

Figure 5.3 Connections between land-use and transportation policy

> The relationship between these two overarching documents provides a unique integrated approach to transport and land-use planning, importantly linked to mutually inclusive and reinforcing implementation processes.

It acknowledges the point made above about Northern Ireland being 'almost totally dependant on a roads-based transportation system' which in turn means that 'an emphasis on the car in the planning of development increases car dependency [which] contributes to social exclusion and reduced accessibility to job opportunities'. Consequently, it outlines a process of integration based on a series of key principles including:

- Land-use allocations should ensure that as far as possible all development sites are accessible by means other than by car, are high-density and mixed-use, and focus on public transport interchanges and areas well served by public transport;
- Transport assessments will provide a review of potential transport impacts of a development proposal; developers' contributions will be sought for the funding of public transport and other infrastructural connections to new development;
- Parking restraints will be encouraged, including the introduction of 'areas of parking restraint' where normal car parking standards are reduced to facilitate objectives such as densification or regeneration;
- High quality 'park and ride' and 'park and share' schemes will be facilitated at public transport interchanges; and,
- A maximum walking distance of 800 metres between new housing developments and the nearest railway station will be established where possible.

Draft PPS 13 should also be considered in association with Draft PPS 12 relating to Housing in Settlements (DRD, 2002c), which focuses on the achievement of higher residential densities, mixed-use developments, brownland locations and, inevitably, the 'integration of residential development with public transport and modes of transport other than the private car'.

It would appear, therefore, that considerable progress has been made in integrating land-use and transportation planning. The evolving policy process has given greater recognition to the relationship between land-use and transportation and the need for an integrated approach, framed within a sustainable development agenda. At the level of generality, sustainability imperatives seem unquestionable (Owen, 1996); however, at the level of implementation sustainable development becomes a contested concept. This is demonstrated by insights developed from an EPSRC project that provided analysis of travel behaviour within the BMA, and the authors' involvement as facilitators for the public consultation process for the Belfast Metropolitan Plan, outlined below.

Travel Behaviour in the Belfast Metropolitan Area

The research on which this section draws was conducted as part of an EPSRC Sustainable Cities project entitled 'Tools for Assessing Consumer, Business and Developer Responses to Sustainable Development Initiatives', and focused on two regional capitals, Belfast and Edinburgh, as case studies. The research aimed to develop new or refine existing planning tools, analysing how travel demand, residential location and business location respond to sustainable development initiatives. This section focuses on a lifestyles-based household survey undertaken in Belfast (and mirrored in Edinburgh) to establish a point of reference for measuring contemporary lifestyles in the city region. The household survey was undertaken in June 2001 and approximately 1,000 questionnaires were collected from 40 wards in the metropolitan area. The household survey contained a vast array of revealed preference housing and transport variables, including migration decisions (e.g. previous, likely future), vehicle ownership (e.g. car, bicycle), parking provision (e.g. home, work) and modal choice (journeys to work, school and main food shopping). A summary of the household survey sample is contained in Table 5.1.

The results of the household survey reinforced the view of Belfast as a car dependant city, with the private car dominating modal choice for journeys to work, school and the main food shopping. For example, the vast majority of the sample in Belfast travelled to work by car (80 per cent) with walking and bus accounting for 4 per cent each. This compares to 60 per cent of the Edinburgh sample using the car to work, with the difference taken up by people walking and taking the bus (13 per cent and 20 per cent respectively). Consolidating the modal choice for travel to work in Belfast is the availability of workplace car parking and a relatively weak city centre in terms of an employment base. Most car and motorcycle users on the journey to work park at the workplace (63 per cent), while a further 20 per cent park in a private car park not available to the public. The majority of this parking is provided by employers (or employment providers) (82 per cent), which tends to be free for employees in most cases (89 per cent). Surveys carried out for the Belfast Alternative Urban Transport Technologies Study in the early 1990s (JUTLU, 1991) showed Belfast also to be unusually generous, as compared with similar sized British cities such as Leicester and Newcastle, in relation to car-parking provision standards for new developments in the city centre. At this time, obviously, attracting new development into a 'troubles-affected' city centre was a priority, so developers could bargain successfully for generous standards.

Given this ease of workplace parking, it is little surprise that for those travelling by car, the primary reasons stated for not using public transport were that it was inconvenient (60 per cent) and that public transport takes too long (43 per cent). Commuting patterns demonstrate that Belfast has a significantly weaker city centre in terms of employment concentration (18 per cent of those employed) than Edinburgh (24 per cent), reflecting Belfast's dispersed

Table 5.1 A summary of the household survey sample

Category	Group	Number	Percent
Gender	Male	494	50%
	Female	490	50%
	Total	984	
Age	18-24	24	2%
	25-34	165	17%
	35-44	194	20%
	45-54	199	20%
	55-64	208	21%
	65-74	142	14%
	74 and over	53	5%
	Total	985	
Household formation	1 adult, no children	110	12%
	2 adults, no children	329	35%
	3+ adults, no children	147	16%
	1 adult, children	52	6%
	2 adults, children	253	27%
	3+ adults, children	60	6%
	Total	951	
House type	Terraced house	264	27%
	Semi-detached house	302	31%
	Detached house	397	40%
	Apartment or flat	7	1%
	Other	13	1%
	Total	983	
House tenure	Owner Occupier	838	85%
	Rented property	146	15%
	Total	984	
Property location	Inner BMA	498	50%
	Outer BMA	497	50%
	Total	984	
Total sample		**995**	

geography of employment (for example, the 1990s witnessed a proliferation of small office development south of the city centre with the availability of free on and off-street car parking). In addition, the Regional Development Strategy promotes the concept of Strategic Employment Locations – large industrial sites of up to 40 hectares, for the purpose of attracting and accommodating major inward investment projects. Within the BMA, it is likely that potential Strategic Employment Locations will be located on greenfield sites at the edge of the Belfast Urban Area, for example in Newtownabbey. However, it is not clear if investment in these sites will be matched by improved accessibility by public transport. The car dependant lifestyle of Belfast residents is further illustrated by modal choice for travel for household food shopping and for children's journey to school/college. The main food shopping is dominated by car travel in Belfast (84 per cent of households) – compared to 75 per cent in Edinburgh. The largest market share was captured by 'Forestside' shopping centre on the edge of the city, which has over 1,500 car parking spaces. Approximately, a third of the Belfast sample had children in full-time education (a total of 638 children). In order of popularity, the mode used for travel to school/college was car (37 per cent), walk (34 per cent) and bus (23 per cent).

Reinforcing these travel behaviour trends are residential preferences and migration within the city. Analysis of migration patterns was undertaken by examining ten spatial sub areas within the BMA. Respondents were given the opportunity to state where they would like to live within the metropolitan area, and for the most part it appears that the desirable places to live are widely dispersed – however, a city centre location was not considered a desirable residential location. A comparison was also made between the location of the previous and current property of households in the survey. The overall trend was one of greater out-migration than in-migration. Movements from inside the city to the city surrounds are 18 per cent, with an in-migration level of 4 per cent, with the remainder moving within inner or outer areas. The two most popular reasons for moving to the current property are cost (44 per cent) and liking the local area (68 per cent). However, related to transport, very few state 'reducing travel costs' as a reason for moving to their current property (8 per cent), while only 6 per cent stated that being too far from work was a reason for moving from a previous property.

These findings present a number of challenges to policy-makers in the Belfast city-region. Although policy appears to embrace an integrated agenda, the emerging geography of Belfast suggests a further dislocation of land-use and transportation. This is particularly the case in relation to employment location and public transport provision. The present dispersed pattern of employment location is difficult to serve effectively and efficiently by public transport as journey-to-work patterns become increasingly lengthy and complex, compounded by changing patterns of work such as 24 hour call-centre offices. A generous supply of free workplace car parking has further encouraged the dominance of the car in travel to work modal choice. The lifestyle and residential preferences of many Belfast residents is highly dependent on the private car, suggesting a considerable obstacle in embracing more sustainable patterns of development, such as promoting higher residential densities with reduced car parking or city centre living. Indeed, community opposition to sustainable development initiatives (albeit combined with support for generalised sustainability principles) was a prevalent feature of the consultation process for the Belfast Metropolitan Area Plan, and this is discussed below.

Sustainability and Community Consultation

The consultation process for the BMAP consisted of a two-stage series of public meetings (information sessions followed some weeks later by area-based debating sessions) supplemented by focus group discussions with disadvantaged groups and integrated seminars with cross-sectoral representation. It was based on the BMAP Issues Paper (DoE, 2001), took place in the first half of 2002 and culminated in a report to government (Gaffikin et al., 2003), which summarised the key findings and reported detailed feedback on an area-by-area basis. This consultation is now the basis of the preparation of the draft plan that will be published in 2004. It was an unusually comprehensive exercise for an Area Plan, and was carried out by a similar consortium of academic and professional experts, as was the case for the Regional Development Strategy.

Sustainable development was one of the 'guiding principles' on which the BMAP Issues Paper (DoE, 2001) was based – it refers to the UK's government commitment to the principle and to its widened definition to include social and economic objectives as well as prudent use of resources and effective protection of the environment. More particularly, it refers to the RDS's commitment to a 'compact metropolitan area' with 'an enhanced quality of urban environment' in which key objectives will be 'the location of new development to reinforce better integration between land use and transportation' and 'the development of a modern integrated and inclusive transport system'.

In the consultation process it emerged that there was a generalised support for the principles of sustainable development and for the objective of better integration of planning and transportation planning, although much scepticism about the achievability of both. This reaction was influenced by the effective participation in many public meetings of environmental pressure groups such as Friends of the Earth and the Belfast Metropolitan Residents Group, who were concerned with green belt protection, urban densification and better public transport. However, this support was tempered with concern about the need to avoid 'town cramming', the need to respect local residential identities and housing quality, and the importance of retaining existing urban 'greenfields' such as parks, golf courses etc. It was also tempered by scepticism about definitions of 'greenfield and brownfield' and about the accuracy and timeliness of the 'urban capacity studies' which were to be the cornerstone of land allocation policy.

However, when it came to area-specific debates, many of the implementation realities of sustainability objectives were seriously questioned. In 'middle-class' areas such as Malone Road and Jordanstown there was concern about developer opportunism in the densification of sites, as well as about the environmental consequences of their operations – reduced privacy, increased traffic and noise, multi-occupancy, student colonisation etc. – not to mention the loss of traditional architectural quality in many instances. In 'working-class' areas in West Belfast, high-density development had unfortunate remembered associations with unpopular projects of the 1960s such as Divis Flats, and there was concern in the Shankhill Road, for example, that the Community Association's objective of attracting middle-class residents back to the area would be undermined if no perceived 'middle-class' (i.e. low-density detached) housing was becoming available. As the report observes (Gaffikin et al., 2003, p.34):

> …while many appreciated the virtues of a compact city, there was a slight tendency to see brownfield development as suitable for somebody else – the towns and villages passing the baton to the main urban area, the suburbs to the city and the more affluent Belfast communities to the inner city.

What is urgently required, the report concludes, is a series of positive architectural role models demonstrating that high-density does not necessarily equate to low quality in terms of housing design. The most obvious place for such experimentation, it was generally agreed, and for the location of significant high-density residential development was the new Titanic Quarter in Belfast's harbour estate, but even here there is some reluctance by the owners to commit to large-scale residential development, and there are limitations in relation to its public transport accessibility.

On the question of public transport there was unanimity – Belfast was perceived to have a poor system as compared with most European cities, and there was much need for improvement. Indeed, a radical agenda is suggested in the report, which would include free public transport, a metropolitan light rapid transit system, congestion charges, integrated travel tickets, flexi-hours for schools, infrastructure-taxes for developers and reductions in the availability of civil-service occupational parking in the city centre.

Matching this wish-list to the draft proposals of the BMTP (DRD, 2003) discussed at a transport consultation conference in February 2003, is an interesting exercise, in that the latter is very firmly grounded in financial reality, although its operational plan (to 2015) is located within a 2025 'vision' which accords with the RDS, in keeping with the rhetoric of land use/transportation integration. Of the above list, affordable rather than free public transport is the objective, and congestion charges 'will be kept under review'. 'Controlled parking areas' and 'restricted parking for new developments' are to be considered, as well as 'integrated ticketing' and 'variable message signing' as part of a campaign to change travel attitudes. Of most interest, however, is the 'commencement of a rapid transit network' focused on four corridors into central

Belfast (from the east, south, west and the Harbour Estate/City Airport, illustrated in Figure 5.2), of which the E-route (from the east along a disused heavy railway track) will be an initial pilot-study, using guided-bus technology which may later upgrade to tram-based systems.

In summary, this consultation exercise reinforces the hypothesis outlined above – a rhetorical support for the principles of sustainability and the practice of land use/transportation integration, combined with a selective reluctance to embrace local changes in residential environment or in lifestyle preferences which might facilitate such principles. There is positive evidence also of more imaginative approaches to transport policy and more mutually-supportive systems of land use/transportation thinking, but these are still conditional in their expression and hedged about with the financial and political uncertainties which are typical of the region.

Conclusion

In terms of urban development, Belfast can perhaps be considered as unique due to the impact of the political conflict in Northern Ireland on urban form, such as persistent ethno-religious residential segregation. However, other trends identified suggests that Belfast has also experienced similar patterns of development to other medium-sized UK and European post-industrial cities – for example, despite the presence of a greenbelt and urban containment policies since 1963, Belfast has experienced processes of metropolitan decentralisation and deconcentration. Therefore, lessons developed from Belfast's recent attempts to integrate land-use and transport policies can provide useful insights for planning practice elsewhere.

In Northern Ireland and Belfast, following the rhetoric of UK planning practice, there has been a shift to an urban development policy discourse that favours a compact city approach, higher residential densities and mixed-use development as a means of reducing the need to travel by car and promoting sustainable development. In this regard much progress has been achieved. A significant development has been the adoption in the Regional Development Strategy of a target of 60 per cent of new housing to be located within brownfield areas. This represents a sea-change in urban development policy as previously less than 30 per cent of new housing was brownfield development. Progress has also been considerable in terms of integrating land-use and transport policy, both at a regional and metropolitan scale. The parallel and inter-linked formulation of a regional spatial strategy and transport strategy is a positive acknowledgement of the inter-dependant relationship between these two functions, particularly within the context of sustainability, and this policy convergence continues at the metropolitan level. Furthermore, Draft Planning Policy Statement 13 is significant in its direct analysis of the car-dependence problem in Northern Ireland and in the appropriateness of its key principles towards achieving more sustainable developments. Initiatives such as land-use allocations based on public-transport accessibility, parking restraint areas free from excessive car-parking requirements and the direct linkage of rail accessibility to new housing development will have a serious impact on both planners' and house-builders' approaches.

However, considerable challenges remain which may lead to difficulties in translating the sustainable development rhetoric into implementation. One concern identified during the household survey is the deep-rooted car dependency culture prevalent in the metropolitan area. Modal choice for journeys to work, school and for shopping were clearly dominated by the motorcar, often facilitated by the availability of free car-parking. Belfast's dispersed geography of employment is difficult to serve effectively and efficiently by public transport as journey-to-work patterns become increasingly lengthy and complex. This is likely to be compounded by the location of Strategic Employment Sites identified in the Regional Development Strategy, which appeared to be selected on the basis of proximity to motorways, rather than access to existing public transport nodes or networks. Clearly, innovative policies are required to address the issue of car-dependency in the city. In particular, car-parking policy needs to be fully integrated into transport demand management as a tool for sustainable urban management, and schemes are needed for linking employment centres to the existing public transport network. Further research is also needed to investigate the impact of increasing numbers of dual-income households on residential preferences and travel-to-work patterns. For example, research from US cities (Jarvis, 2003) suggests that intra-household negotiation and compromise on residential location has a significant impact on travel behaviour as households seek a 'hub' residential location with access to large labour markets.

Although the 'policy/expert' community increasingly favours a more compact, higher density form of urban development, a number of tensions can be identified, particularly in relation to increasing residential densities to accommodate future housing demand. Firstly, it is not clear if private house-builders have 'bought into' this new vision for housing development. The house-building lobby vigorously opposed the 60 per cent regional target for brown field housing during the preparation of the RDS (see Neill and Gordon, 2001), preferring to maintain the existing pattern of greenfield development. Secondly, during the BMAP public consultation process, increasing residential densities proved perhaps the most contentious issue at community meetings, both within low-income and middle-high income areas. This suggests that the identification of potential locations in the plan process and proposals for high-density housing development will be opposed by organised residents' groups. A third issue relates to consumer or lifestyle choice for potential house-buyers. For example, if a compact, high-density city is rejected by consumers in favour of lower density housing developments, it is possible that locations and settlements beyond the greenbelt could prove popular. In this case, residential preference for lower densities would lead to a more car dependent pattern of development as people move to locations outside of the metropolitan area. Reinforcing this perspective, it was apparent from analysis of the household survey that the city centre as a residential location appears to have only a limited appeal.

In this context, positive urban design models are required to demonstrate to both house-builders and consumers that higher-density residential development can satisfy contemporary lifestyles at various stages of the lifecycle, including suitability for families, and also to counter the poor image of city centre living. In this sense, quality urban design can help to address the emerging gap between the 'policy/expert' community and individual aspirations. In addition, it may be appropriate to consider marked-based policy instruments to encourage both higher densities and city centre residential development. For example, in the Republic of Ireland in the late 1980s and early 1990s tax incentives were successfully employed to encourage both developers and consumers to invest in inner city areas. In common with many planning initiatives, the rhetoric which characterises recent strategic planning and transport documents in Northern Ireland is consistently positive in relation to issues of sustainability, although its translation into reality will probably be problematic. Transport and land-use planners in Northern Ireland have generally been cautious in their promises as well as limited in their ambitions. The recent break from this tradition, regardless of the long-term realities, is therefore a welcome innovation. Nevertheless, the difficulties experienced in policy integration identified, suggest a considerable challenge to implementing a sustainable development agenda.

Notes

[1] This research was undertaken as part of an EPSRC Sustainable Cities project entitled 'Tools for Assessing Consumer, Business and Developer Responses to Sustainable Development Initiatives'. The project was undertaken jointly by the Transport Research Institute, Napier University and the School of Environmental Planning, Queen's University Belfast, and staff involved in the project were Austin Smyth, Tim Ryley and James Cooper (all Napier University) and Malachy McEldowney, Geraint Ellis and Mark Scott (Queen's University).

[2] The public consultation process for the Belfast Metropolitan Area Plan was undertaken by a consortium of the School of Environmental Planning, Queen's University Belfast and Price Waterhouse Coopers. Queen's University staff involved in the exercise were Frank Gaffikin, Malachy McEldowney, Ken Sterrett, Stephen McKay, Jayne Bassett and Mark Scott.

References

Banister, D. (1997) 'Reducing the need to travel', *Environment and Planning B: Planning and Design*, Vol. 24, pp. 437-449.

Banister, D. (1998) *Transport Policy and the Environment*, London: E&FN Spon.
Banister, D. (1999) 'Planning more to travel less: land use and transport', *Town Planning Review*, Vol. 70, pp. 313-338.
Breheny, M. (1992) 'The Contradictions of the Compact City: A Review', in Breheny, M. (ed.) *Sustainable Development and Urban Form*, London: Pion.
Burton, E. (2003) 'Housing for an Urban Renaissance: Implications for Social Equity', *Housing Studies*, Vol. 18, pp. 537-562.
Commission of the European Communities (1990) *Green Paper on the Urban Environment*, Brussels, CEC.
Cooper, J., Ryley, T. and Smyth, A. (2001) 'Contemporary lifestyles and the implications for sustainable development policy: lessons from the UK's most car dependant city, Belfast', *Cities*, Vol. 18, pp. 103-113.
DOE (Department of Environment (2001) *Belfast Metropolitan Area Plan 2015 Issues Paper,* Belfast: DOE.
DOE (Department of Environment) (2003) *The Public Consultation on the Belfast Metropolitan Plan (BMAP)* 2015, Belfast: DOE.
DOELG (Department of Environment and Local Government) (1999) *Residential Density: Guidelines for Planning Authorities*, Dublin: Stationery Office.
DOELG (Department of Environment and Local Government) (2002) *The National Spatial Strategy 2002-2020*, Dublin: Stationery Office.
DRD (Department of Regional Development (2001) *Shaping Our Future - Regional Development Strategy for Northern Ireland 2025*, Belfast: DRD.
DRD (Department of Regional Development) (2002a) *Proposed Regional Transportation Strategy for Northern Ireland*, Belfast: DRD.
DRD (Department of Regional Development) (2002b) *Draft Planning Policy Statement (PPS) 13, Transportation and Land-use*, Belfast: DRD.
DRD (Department of Regional Development) (2002c) *Draft Planning Policy Statement (PPS) 12, Housing in Settlements*, Belfast: DRD.
DRD (Department of Regional Development) (2003) *Belfast Metropolitan Transport Plan Conference Papers*, Belfast: DRD.
Ellis, G. and McKay, S. (2000) 'City Management profile, Belfast', *Cities*, Vol. 17, pp. 47-54.
Gaffikin, F. and Morrissey, M. (1999) 'Sustainable Cities', in Gaffikin, F. and Morrissey, M. (eds) *City Visions, Imaging Place, Enfranchising People*, London: Pluto Press.
Gaffikin, F., McEldowney, M. Sterrett, K. and McDonagh, P. (2003) *The Public Consultation on the Belfast Metropolitan Area Plan,* Belfast: DOE.
Hass-Klau, C. (1990) *The Pedestrian and City Traffic*, London, Belhaven.
Heath, T. (2001) 'Revitalizing Cities, Attitudes towards City-Center Living in the United Kingdom', *Journal of Planning Education and Research*, Vol. 20, pp. 464-475.
Jarvis, H. (2003) 'Dispelling the Myth that Preference make Practice in Residential Location and Transport Behaviour', *Housing Studies*, Vol. 18, pp. 587-606.

JUTLU (Joint Urban Transport and Land-use Unit) (1991) *Belfast Alternative Urban Transport Technologies Study*, Belfast: Queen's University / University of Ulster.

McEldowney, M. and Sterrett, K. (2001) 'Shaping a Regional Vision: the Case of Northern Ireland', *Local Economy*, 16, pp. 38-49.

Neill, B. and Gordon, M. (2001) 'Shaping our Future? The Regional Strategic Framework for Northern Ireland', *Planning Theory and Practice*, Vol. 2, pp. 31-52.

Owen, S. (1996) 'Sustainability and Rural Settlement Planning', *Planning Practice and Research*, Vol. 11, pp. 37-47.

Owen, W. (1972) *The Accessible City*, Washington: Brookings Institution.

Tewdwr-Jones, M. and Williams, R. (2001) *The European Dimension of British Planning*, London: Spon Press.

Urban Task Force (1999) *Urban Renaissance Report*, London: Stationary Office.

Vigar, G. (2002) *The Politics of Mobility: Transport, the Environment and Mobility*, London: Spon Press.

Chapter 6

Improving Energy Efficiency in Urban Areas

J. Peter Clinch

Introduction

During the 1990s, the Irish economy was the fastest growing economy in Europe. The growth of output was unprecedented in an Irish context and the growth in employment a record in an international context. Between 1990 and 2000, Gross Domestic Product doubled with the real growth rate at a high of 11 per cent in 1999. While the growth rate fell back considerably in the first few years of the new millennium, growth continues to be in the order of four to five percent. While this extraordinary economic success, in particular the fall in unemployment, has been most welcome, the pace of the growth has had significant implications for settlement patterns, energy use and the environment.

Energy consumption tends to be strongly related to economic growth. Figure 6.1 compares indices of energy consumption and GDP. Because the growth in GDP is so extraordinary, the growth in energy consumption may not, at first glance, seem so spectacular. However, the data shows that total final consumption of energy has risen by 70 per cent between 1980 and 2000 with the tertiary sector rising more than proportionately. Energy consumption by industry has grown by 21 per cent. The development of 'clean' industry has made this latter figure a relatively good performance when one considers that the growth of industrial output is an order of magnitude greater. Energy consumption in the domestic sector grew by 31 per cent but, most remarkably, it has more than doubled in the transport sector. This suggests that the key sectors of concern in relation to energy use and the environment in urban areas are the transport and residential sectors. The primary objective of the chapter is to examine the environmental implications of energy use in these sectors, to examine the factors that inhibit improved energy performance and to provide some recommendations to improve performance.

Environmental Implications of Energy Consumption

The rapid growth of the Irish economy has engendered activity that has intensified pressure on environmental endowments. Increased energy use has implications for

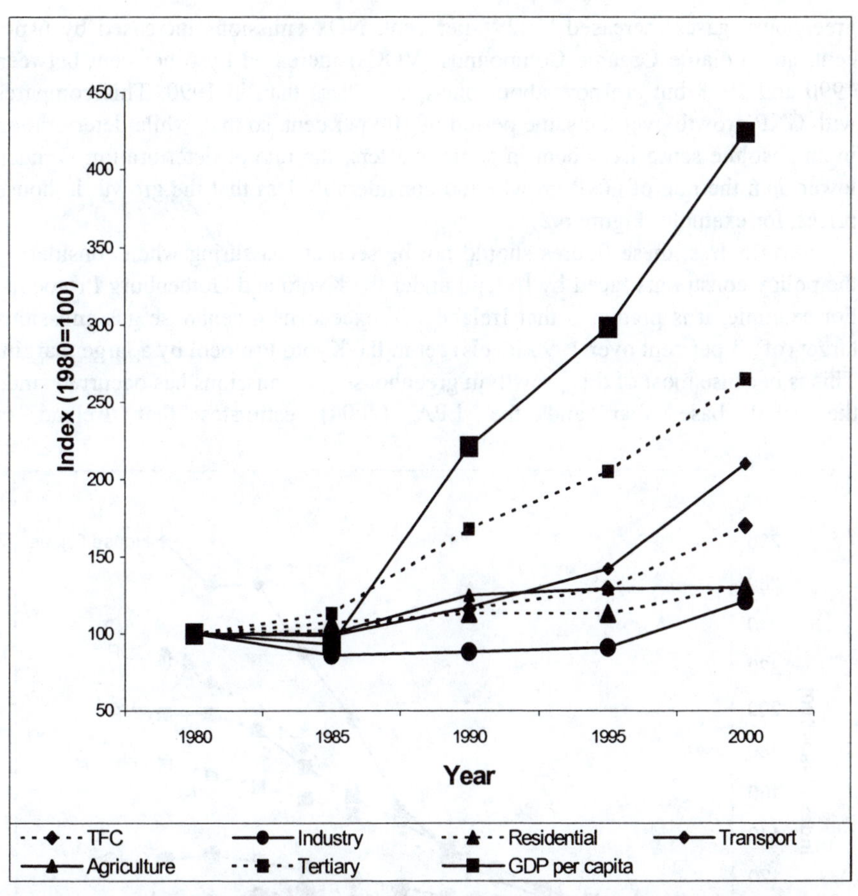

Figure 6.1 Indices of GDP per capita and energy consumption by sector
Source: Healy (2001)

local, regional and global environmental performance. At a global level, the greatest concern surrounds the implication of global warming, thought to result partially from the burning of fossil fuels for energy production. At a regional level, energy production from non-renewables results in the emission of acidification precursors which cause acid rain. At a local level, carbon monoxide (CO) and particulate matter (particularly PM10) from vehicles are known to have serious implications for human health as are some emissions from industry and energy generation. Sulphur dioxide (SO2) also results in the degradation of buildings in urban areas.

Over the Celtic Tiger era, Ireland has had a mixed performance as regards air quality. Between 1990 and 1998, emissions of SO2 stayed relatively constant, but in 2002 was nearly half what it was at the beginning of the 1990s. CO was reduced by 25 per cent between 1990 and 2002. However, emissions of greenhouse gases increased by 29 per cent, NOx emissions increased by 6 per cent, and Volatile Organic Compounds (VOCs) increased by 6 per cent between 1990 and 1998 but are now about one-quarter less than in 1990. This compares with GNP growth over the same period of 100 per cent, so that, while deterioration in an absolute sense is evident in some quarters, the rate of deterioration is much lower than the rate of GNP growth and considerably less that the growth in house prices, for example (Figure 6.2).

Nevertheless, these figures should not be seen as reassuring when considering the policy constraints faced by Ireland under the Kyoto and Gothenburg Protocols. For example, it is predicted that Ireland will exceed its greenhouse gas emissions target (of 13 per cent over 1990 levels) set in the Kyoto Protocol by a large margin. This is because most of this growth in greenhouse gas emissions has occurred since the 1990 base year and the EPA (2004) estimates that Ireland is

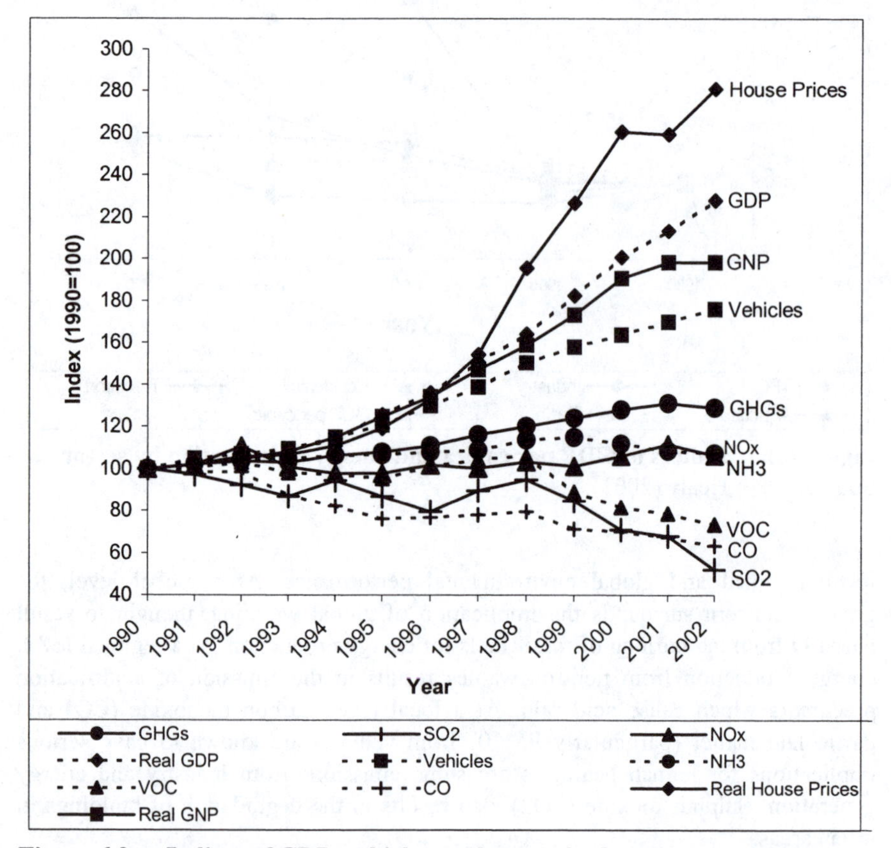

Figure 6.2 Indices of GDP, vehicle numbers and selected air emissions, 1990-2002

already 29 per cent over its Kyoto limit. Ireland's unprecedented growth has resulted in estimates that, under a 'business as usual' scenario, total emissions would increase by between 37 per cent and 41 per cent by 2010 (Figure 6.3).

Although there was a decrease in the first half of the decade, emissions of SO2, NOx, VOCs and CO increased again during the late 1990s due to an increase in consumption of primary fuels. According to Stapleton et al. (2000), the variation in NOx (Figure 6.4) emissions over the decade is the result of a programme to retrofit some of the largest power stations with NOx control technologies which led to some units being off-line for long periods. Ireland failed to meet the targets of 105,000 tonnes of NOx in 1994 (actual was 115,600 tonnes) set by the Sofia Protocol (1988) and it failed to meet the target of 157,000 tonnes of SO2 in 2000 set by the Oslo Protocol (1994). Under the recently signed Gothenburg Protocol (1999) (and a related EU Directive), Ireland is committed to reducing SO2, NOx, NH3 and VOC emissions to 42,000 tonnes (from 96,300 in 2002), 65,000 tonnes (from 125,300 in 2002), 116,000 tonnes (from 118,000 in 2002) and 55,000 tonnes (from 81,400 in 2002) respectively by 2010. With the exception of NH3, achievement of these targets remains a considerable challenge.

Examining national air quality statistics ignores impacts of considerable import for quality of life, and especially urban living. While incomes have risen, urban dwellers face increased road congestion, longer commuting distances,

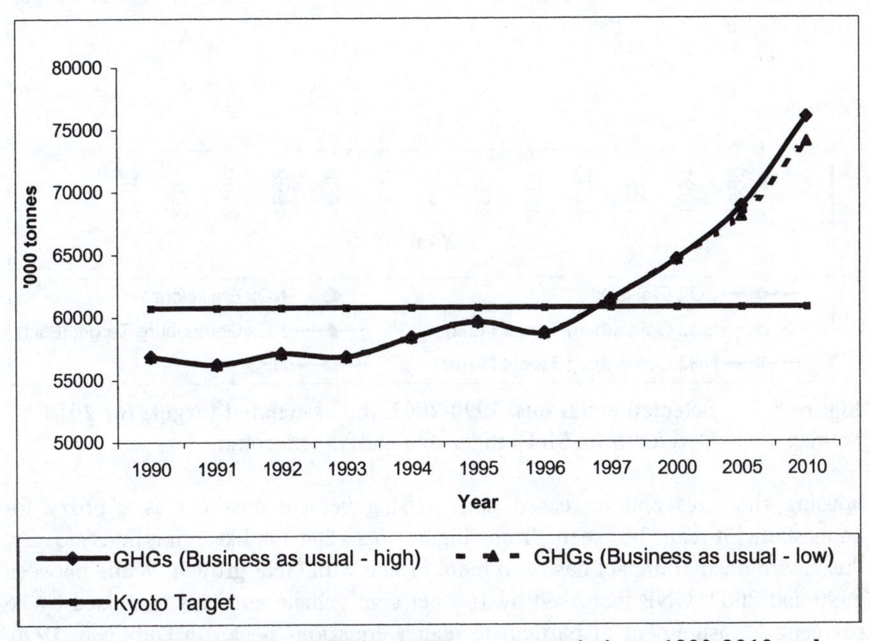

Figure 6.3 'Business as usual' greenhouse gas emissions, 1990-2010 and associated target for 2012

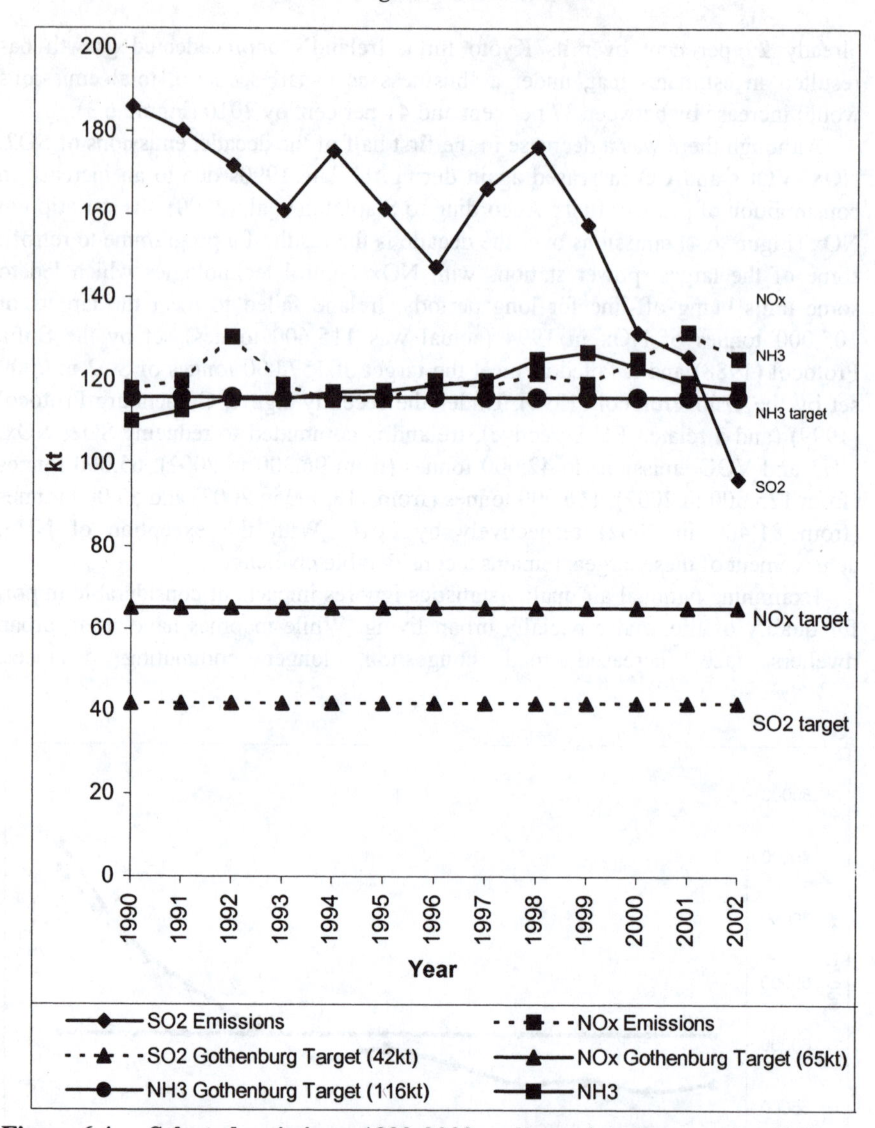

Figure 6.4 Selected emissions, 1990-2002 and associated targets for 2010
Source: Derived from McGettigan and Duffy, pers comm.

housing shortages and increased noise. Using vehicle numbers as a proxy for congestion, it can be seen from Figure 6.2 that, unlike emissions to air, this environmental impact has been more in line with GNP growth. While between 1990 and 2002, GNP increased by 100 per cent vehicle numbers increased by 76 per cent. Measurement of particulate matter emissions began in Dublin in 1996. Emissions have been reduced in the areas of most concern including College Green but the number of days exceeding limit values (50 mg/m3 is in excess of that which will be permitted by EU law in 2005) remains high.

Urban Transport

The rapid growth in disposable incomes in Ireland has contributed to an unprecedented rise in the levels of car ownership and congestion. The number of cars and good vehicles grew by 82 per cent and 63 per cent respectively between 1990 and 2002 (EPA, 2004). Car ownership rates were 370 per 1,000 in 2002 (Figure 6.5) and this has surpassed the figure of 342 per 1,000 originally predicted for 2011. Nevertheless, this is still below the EU average so further growth can be expected. The proportion of people driving a car to work has also

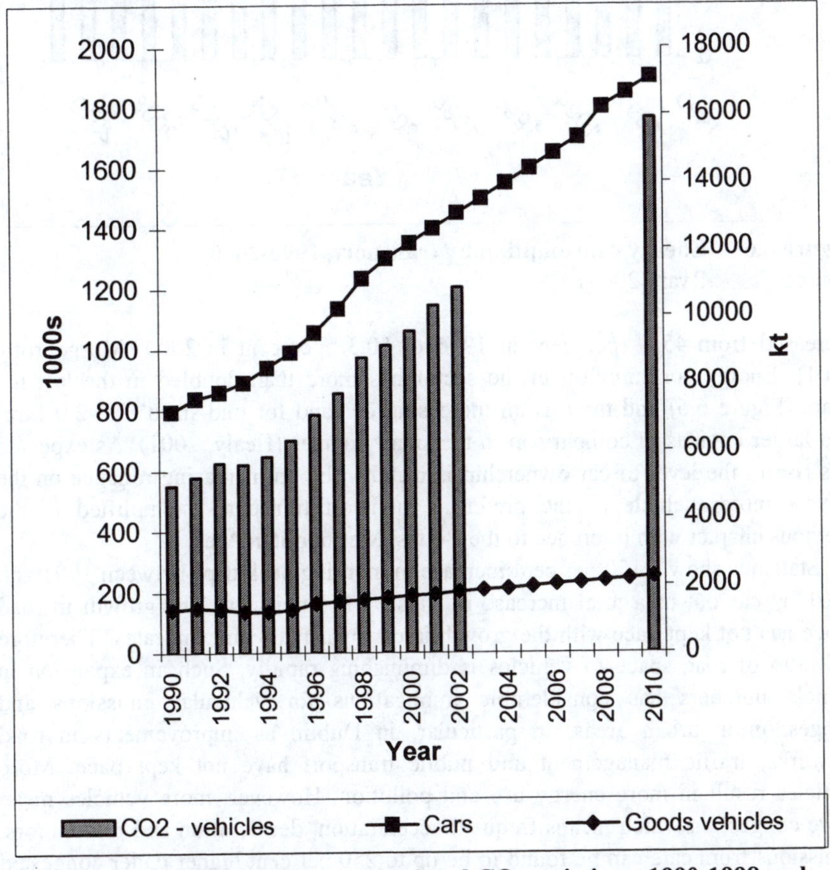

Figure 6.5 Vehicle numbers and related CO_2 emissions, 1990-1998 and forecasts to 2010

Source: Clinch (2002), Ryan (2004)

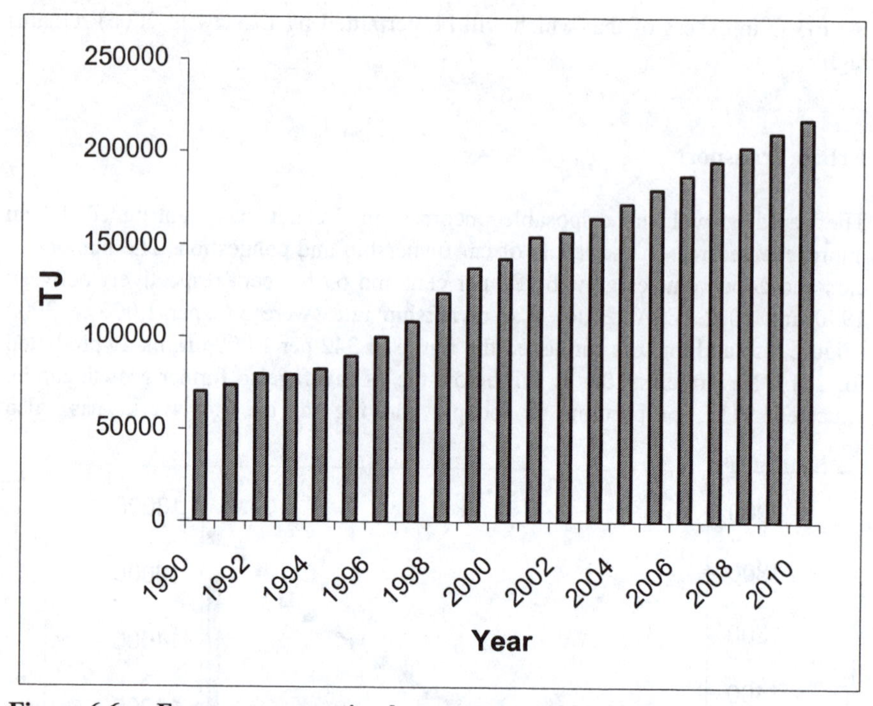

Figure 6.6 Energy consumption by transport, 1990-2010
Source: Ryan (2004)

increased from 45.1 per cent in 1996 to 50.3 per cent in 2000 (Morgenroth, 2001). Energy consumption in the sector has more than doubled in the last ten years (Figure 6.6) and there is an increasing demand for mid-sized (1.3-2.0 litre) and larger engines in comparison to the small engines (Healy, 2001). As expected this rise in the level of car ownership is matched by an increasing reliance on the private motor vehicle as the preferred mode of transport, exemplified in the previous chapter with reference to the Belfast Metropolitan Area.

Statistics show a 150 per cent increase in morning peak trips between 1991 and 2001 by car out of a total increase in trips of 127 per cent. The growth in road space has not kept pace with the growth in car ownership and trip rates. Therefore the ratio of road space to vehicles is diminishing rapidly. Such an expansion in vehicle numbers has considerable implications for vehicular emissions and congestion in urban areas, in particular, in Dublin as improvements in road networks, traffic management and public transport have not kept pace. More vehicles result in more energy use and pollution. However, more vehicles mean more congestion which means frequent acceleration, deceleration and idle motors. Emissions from cars can be found to be up to 250 per cent higher under congested conditions than under free flowing traffic. Convery (2001) notes that the majority of the local air pollution in our cities, and the associated health and other dysfunctions, are a product of emissions from road-based transport. As such, transport is the source of most fine and ultra fine particles, and these are now associated with lung and other health dysfunction (Pope et al., 1995). The transport sector accounts for over 44 per cent of the increase in greenhouse gas emissions (Table 6.1) and over 50 per cent of emissions of NO_x and VOCs. Between 1990 and 2002, the largest increase in greenhouse gas emissions of 127 per cent occurred in the transport sector. The EPA (2004) state that emissions from road traffic are now the primary threat to the air quality of Ireland. The areas with large volumes of traffic (College St. in Dublin and Old Station Rd. in Cork) have the highest levels of particulate matter but limit values are also exceeded in suburban areas such as Rathmines and the Phoenix Park, key routeways in 'suburb-suburb' commuting.

Implications of Settlement Patterns for Transport and Energy Use

Housing policy during the Irish boom was particularly inadequate and the spread of the Dublin commuter belt to include all of Leinster, and even further afield, has led to many undesirable externalities. These include further congestion, increased travel time to work, rising frustrations and stress, increased fuel use and associated greenhouse gas and pollution emissions. Much of the energy in the new Dublin economy is deriving from developments along the 'C ring' motorway circling the city to the west, with nodes at Swords/Dublin airport (Motorola, Hertz), Blanchardstown/Mullhudart (IBM complex), West Dublin/North Kildare (Intel, Hewlett Packard), Tallaght (City west, Park west) and Leopardstown/Sandyford (Microsoft, Cherrywood Technology Park). This development of an 'edge city' is symptomatic of developments in many urban

Table 6.1 Breakdown of greenhouse gas emissions by sector, 1990 and 2010

Sector	*Proportion of Emissions 1990 (per cent)*	*Proportion of Emissions 2002 (per cent)*	*Proportion of Emissions 2010 (per cent)*
Agriculture	34.6	28.5	25.6
Energy	21.6	24.6	25.0
Residential	13.1	9.7	9.0
Transport	9.5	17.0	18.9
Industrial	8.0	8.3	8.0
Process	5.4	4.8	6.6
Commercial/Inst.	4.5	4.5	5.4
Waste	3.3	2.6	1.5

Source: Consultation Group on Greenhouse Gas Emissions Trading (2000), EPA (2004)

areas, but has happened with particular rapidity in the Dublin region. Except for some apartment development in the inner city, most housing is low density suburban, and this process of urban sprawl and dispersal seems to be accelerating, with much of the development leapfrogging the established commuter belt. Between 1997 and 1998, the level of planning permissions for new housing in the outer Leinster counties (60 km+ from Dublin) grew by 40 per cent, compared with a 14.7 per cent growth in the Dublin region, and long distance commuting by rail and bus, but mainly by car, is becoming characteristic (Clinch et al., 2002).

Market Failure and Energy Efficiency in Transport

In this context then, the principal problem in relation to energy use in urban transport and availability of road space reflects quite closely the problems raised in Garrett Hardin's (1968) classic piece on the 'tragedy of the commons'. Once road tax is paid, everyone has access to the road network at all times (unless the road is tolled). In the short-term the supply of road space is relatively fixed and thus, beyond a certain point, marginal congestion costs increase. As more and more cars come onto the roads, the costs imposed by each additional car become greater and greater. If one car stays off the road, the other cars continuing to use it are rewarded by less congestion. Thus, there is a classic free-rider problem: why should I stay off the roads and let others benefit? This is the standard problem in relation to common-property type resources. It explains why urging motorists to leave the car at home (such as on 'car free days') has a very limited effect.

So, what is the solution? The economist's analysis of the situation is that individuals must be made to face a certain proportion of the costs imposed on others by their journey. Note the concept of marginal cost is the key, that is to say, what they pay should relate to the costs of an additional journey rather than a fixed amount at the beginning of the year (such as road tax or insurance). Estimates have shown that private motorists pay in all motor taxes an amount greater than total congestion costs. However, this is not the issue. These fixed charges do not address the problem of congestion – in fact, they could even encourage more frequent car use: Why leave a fully taxed and insured car sitting idle in the driveway? But making individuals pay some of the marginal costs imposed on others by each trip would lead to a more efficient use of road space.

In addition to these so-called 'demand-side' measures, which influence people's demand for road space, supply-side measures are important instruments to alleviate congestion. Such measures alter the supply of different forms of road space and modes of travel. An increase in the availability and/or quality of substitutes to car travel also increase the effectiveness of demand-side instruments to effect a modal shift.

Buildings and the Residential Sector

However altering transportation habits alone will not be sufficiently effective to help meet international responsibilities, given that the demand for energy in the residential sector has increased by about a third since 1980 and this is expected to increase by a further one-quarter over the next decade. This is despite the increasing penetration of domestic energy-efficiency technologies, more efficient heating systems and higher building standards and regulations. McLoughlin (2001) has shown that consumption of coal in the residential sector fell significantly over the 1990s (Figure 6.7). The principal reason for the decline was the introduction in 1990 of a ban on the sale and distribution of bituminous coal in Dublin. This was extended to all other major urban centres in Ireland subsequently and has resulted in residential being the only sector in which greenhouse gas emissions did not rise (EPA, 2004).

Recent population growth and estimates forecasting continuing population increases over the proceeding decade are shaping the future energy-consumption patterns in this sector. In addition, more households are using high-energy-consuming electrical appliances like dishwashers, tumble-dryers, microwave ovens, and so forth, again due to rising living standards and behavioural and comfort choices, and again negating the gains made in overall levels of residential energy efficiency (Healy, 2001).

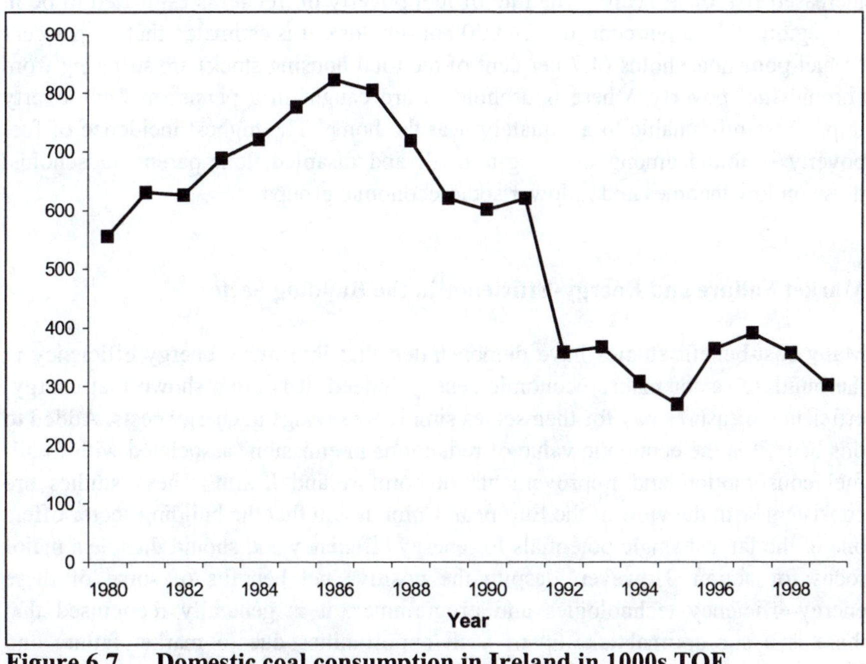

Figure 6.7 Domestic coal consumption in Ireland in 1000s TOE
Source: McLoughlin (2001)

Residential Energy-Efficiency and Fuel Poverty

The consequences of poor energy-efficiency in the residential sector are not only environmental. Despite enduring relatively mild winters, Ireland and the UK have the highest rates of seasonal mortality in northern Europe, and it has been shown that such mortality rates result, in no small part, from the inadequately protected, thermally-inefficient housing stocks in these countries (Clinch and Healy, 2000; Curwen, 1991). There are also strong associations between inadequately heated homes and increased rates of morbidity; higher incidences of various cardiovascular and respiratory diseases have been associated with chronic cold exposure from within the home through living in fuel-poor conditions (Collins, 1986; Evans et al., 2000). Thus, when temperatures fall during a typical British or Irish winter, households need to increase their expenditure on fuel considerably to heat their home adequately, owing to the poor level of heat retention in their dwellings. The problem of fuel poverty occurs, therefore, when a household does not have the adequate financial resources to meet these winter home-heating costs, and because the dwelling's heating system and insulation levels prove to be inadequate for achieving affordable household warmth.

Healy and Clinch (2004) show that, while the penetration of lagging jackets, double glazing and central heating have improved substantially over the past five years, the Irish housing stock remains considerably under-protected from the outdoor environment, leaving the vulnerable in society open to fuel poverty and increased risk of ill health. The rate of fuel poverty in Ireland is estimated to be in the region of 17.5 per cent (or 226,000 households). It is estimated that 27 per cent of fuel-poor households (4.7 per cent of the total housing stock) are suffering from chronic fuel poverty, where householders are caught in a persistent fuel-poverty trap, constantly unable to adequately heat the home. The highest incidence of fuel poverty is found among the long-term ill and disabled, lone-parent households, those on low incomes and in lower socio-economic groups.

Market Failure and Energy-Efficiency in the Building Sector

Many cost-benefit studies have demonstrated that improving energy efficiency in the building sector makes economic sense . Indeed, it is often shown that energy-efficiency measures pay for themselves simply by savings in energy costs. Added to this benefit is the economic value of reductions in emissions associated with fossil-fuel consumption and improvements in comfort and health. These studies are consistent with the view of the European Commission that the building sector offers one of the largest single potentials for energy efficiency and should thus be a major focus for action. However, despite the positive net benefits of some of these energy-efficiency technologies and programmes, it is generally recognised that there is a sub-optimal take-up of such opportunities due to market failure and government failure which result in a number of problems. Firstly, private and social benefits and costs may differ. External benefits which are captured by wider society (e.g. reductions in environmental emissions) may not to be considered when a private individual is considering whether to invest in such measures. The payback periods and net benefits of various measures and programmes are adversely affected by the exclusion of non-private benefits. In addition, while some of the benefits may be private in nature, they may not be recognised or considered by those who benefit. Energy savings may well be considered but improvements in health, being non-monetary in nature, are often not known about or recognised when making financial decisions.

The consideration of time is usually of considerable importance when assessing the net benefits of energy-efficiency technologies or programmes. Those who are considering improving the energy-efficiency characteristics of a building are likely to carry out a financial analysis. The market interest rate is likely to be used in these calculations as it reflects the opportunity cost of capital. These rates may be somewhere in the region of 10 per cent which would reduce the net present value of future energy savings and thereby increase the payback period and possibly result in a negative return. In addition, those who make the decisions may not reap the rewards. There may be market inefficiencies as regards the incentives for developers to adopt improved technologies. This can result because those making the decision as to whether to upgrade the energy-efficiency standards of a new building may not be the occupiers of the completed building. It may well be that those who make the decisions regarding whether to install the better technologies or whether to rent or purchase a building which embodies these technologies, may not be those who occupy the building day-to-day. Tenants are not generally responsible for the energy-efficiency standards of the buildings they occupy. This is also a particular problem in the domestic / household sector. For example, some of the least energy-efficient houses in the UK and Ireland are tenant-occupied. Tenants may feel that they are not responsible for undertaking investments in energy efficiency or authorised to do so.

Socio-economic considerations also play an important role in relation take up of energy efficiency opportunities in the household / domestic sector. The least energy-efficient households are more likely to be lower income households . Such households are much less likely to have available funds and, thus, are most likely to have to resort to a loan. They are less likely to be in the position of accessing credit (particularly at the market rate of interest) and they are more likely to have more pressing alternative uses for any extra funds. They may, additionally, have an aversion to borrowing funds, as has been reported by Salvage (1992). It has also been shown that low-income households tend to have higher discount rates, i.e. they exhibit myopic tendencies whereby they place a greater value on income now as opposed to in the future, partly resulting from the higher degree of uncertainty about the future stemming from their financial instability.

There may be considerable information asymmetries. If the market worked effectively, the monetary value of the energy-efficiency measures would be reflected in the value of the buildings and this would provide an incentive for the technologies to be implemented. Information asymmetries inhibit this function of the price mechanism. Closely related to the information problem is that of the fixed costs of learning about, and administering, energy-conservation measures. Examples of transactions costs include the time agents must spend to learn about the various options, oversee the work, and deal with any disruption.

Policy Measures to Promote Energy Efficiency: The Transport Sector

As has already been suggested, the major growth in energy demand and associated environmental emissions and congestion have occurred in the Greater Dublin Area so there is, by necessity, a concentration on this area with regard to policy recommendations although many would apply to other large urban areas such as Cork. Transport policy instruments can be categorised as supply-side and demand-side measures.

Supply-Side Measures

Control centres (such as in Dublin City Council Dublin) that make the best use of the supply of road space in urban areas can make a major contribution to reducing congestion and thereby reducing environmental emissions. These centres can facilitate rapid reaction to any incident that is hindering the flow of traffic, whether it is an accident or an illegally parked car. VMS Parking Data provide motorists with information on parking availability in multi-storey car parks if signs are positioned at key intersections and thereby help to direct the flow of traffic searching for parking.

Appropriate management of roadworks and extensive development of cycle lanes throughout cities is underway. For examples, when completed, the total length of cycle lanes in Dublin will amount to 160km. In addition, there are a total of 11 'quality bus corridors' (dedicated bus lanes that operate from 7am to 7pm with traffic signal priorities). However, the continuation of cash fares has led to the observation of bunching of buses on QBCs. On the most successful of the bus routes implemented to date, there has been a 190 per cent increase in passenger levels during the morning peak. This level of increase would indicate that the programme has, to some extent, been successful in triggering some modal shift from low to high occupancy vehicles in the last couple of years along those routes. There is a concern regarding the distributional impacts of the QBC policy with a suggestion that service on the QBC routes has been kept up at the expense of other routes. Also, park and ride facilities are completely inadequate and result in considerable costs to those who live in areas close to QBCs who have to tolerate people parking on kerbs.

With respect to access, public transport will always be at a disadvantage when competing with a private mode of transport for reliability and convenience. A private car will always be there, ready to leave when you are. It is a common complaint about the bus services in Dublin that they are simply too unreliable and therefore not to be trusted as a means of travel. Plans however are under way to provide real time passenger information at bus stops by 2005/6, detailing when the next bus will arrive. There is still a need for improved connectivity between all modes of public transport and a fully integrated ticketing system that reduces the delay (and consequent convoys of buses) caused by cash fares.

Welcome developments have been the deregulation of the taxi markets, which has helped to reduce queuing times, the proposal to open up public transport to competitive tender and the opening of both lines of the Luas light rail system (although questions remain as to how effective this service will be at encouraging modal shift and whether, as an investment, it represents good value for money and whether joining the two lines is feasible and desirable). The opening of the Dublin Port Tunnel, which will divert heavy traffic away from Dublin city centre will be of great benefit.

Much work has also been put into making the city more pedestrian friendly. Measures taken include: the installation of more pedestrian crossing facilities, increasing 'Green man' time, redesign of junctions with pedestrian safety in mind, and timers that countdown to the next crossing to improve pedestrian patience. However, these improvements in the provisions for pedestrians come at the expense of traffic flow.

The potential exists at present to provide commuters with a great deal of information to improve flow through the transport network and minimise waiting time via advanced commuter information systems. As wireless communication advances and wireless devices saturate the market, it is likely that a wide range of information from 'When is the next bus to town?' to 'Where can I find an on-street parking space?' will become available to commuters at the touch of a button. The increased efficiency and reliability that will come with this will certainly improve the attractiveness of public transport (Clinch and Kelly, 2002).

Specific details regarding transport planning in the Greater Dublin Area are detailed in the Dublin Transportation Offices 'A Platform for Change 2000-2016' (DTO, 2001) and on the DTO's website, www.dto.ie.

Demand-Side Measures

On the other hand, fiscal measures include excise duty on motor fuels, sales tax on vehicles (VRT), annual road tax and tax on commuting expenses. The sales tax rate and the annual road tax on private vehicles depend on engine size. These were introduced as luxury taxes rather than for environmental purposes and do not take into account engine characteristics such as emissions and noise. However there are incentives for purchasers of hybrid (battery and petrol) motor vehicles although such cars are still relatively expensive. In urban areas the size of the vehicle is probably more important as an indicator of social costs than engine capacity. The exponential growth of large 'all terrain vehicles' is generating an increasingly severe problem on the narrow streets of the city. The tax rate ought to take account of the vehicles dimensions in order to discourage the wasteful use of scare urban road space.

Parking policy is one of the most potent tools at the disposal of city councils for influencing modal choice and a significant source of revenue. The pricing of parking is a second-best approach for charging the private motorist for the marginal costs imposed by their travel. The power of this tool depends on the proportion of spaces under the control of the policymaker. The success of this tool will be greatly enhanced if private non-residential spaces (PNR) are covered by the policy, which is not the case at present.

A government commissioned report recommended a road-pricing charge of €3.81 per car entering Dublin city centre during peak hours and estimated that the system would pay for itself within 1 year. However, the recommendations have not been adopted and there is no use of congestion charging at present. The best examples of implemented road pricing schemes come from Singapore and London. However, it is notable that these cities have exemplary public transport systems that include a metros. Dublin, for example, does not and, therefore, the sensitivity of demand for roadspace to the price is likely to be substantially lower.

Policy Measures to Promote Energy Efficiency: The Residential Sector

There are a number of instruments available to policy-makers to correct for market failure in the residential sector. Regulation, also known as command-and-control, endeavours to improve the performance of the market via the setting of standards e.g. building regulations. Non-compliance with a standard results in a penalty, usually in the form of legal action and/or fines. Regulation is likely to be most effective for new buildings where minimum standards can be set. The energy rating system to be applied to all buildings (under the EU Energy Performance Directive) is a significant step in improving the energy efficiency of buildings.

Sustainable Energy Ireland plays a key role in information provision via an information service and media advertising. This assists in bringing energy efficiency opportunities to the attention of those households who, once they are aware of the opportunities, will find it worth their while to invest in energy efficiency. However, further investment in research and development into the most appropriate technologies and the most effective points of intervention is required. In addition, information on the energy performance of the built environment is extremely patchy.

In addition to an information campaign, it is equally important to minimise transactions' costs. Those households with relatively high incomes may also have a high opportunity cost of time, i.e. they may be unwilling to exchange a relatively small saving in energy expenditure for the time involved in sourcing a company to carry out work and arranging for them to spend time in their house. A scheme which targets particular areas of housing for retrofitting and lines up appropriate construction firms can minimise these costs. Such large-scale schemes can capture significant economies of scale.

Emissions Trading is another market-based instrument. Rather than being a price instrument (like a tax), it is a quantity-based instrument. The Kyoto Global Warming Protocol contains a provision to allow for such trading. Compliance with the greenhouse emission quotas can be achieved, in part, by purchasing from others who have a quota to spare. A price emerges for the permits which reflects the scarcity value of the environment. The European Union Emissions Trading Scheme will commence in 2005 and Ireland has already allocated initial quotas. However, the transport and residential sectors are not included directly.

Environmental taxes and charges provide a much more effective approach to providing incentives for small pollution sources (such as households) and mobile sources (such as vehicles). These instruments are put in place by a policy-maker to alter market signals to encourage or discourage certain activities or behaviour. A tax on energy generated from fossil fuels was part of the government's Climate Change Strategy to reduce emissions of greenhouse gases by providing incentives to invest in energy-conservation measures. However, the Irish government recently abandoned plans to introduce such a tax.

Other instruments include the removal of subsidies, if any, on energy products and tax relief and grants for energy-conservation measures. Voluntary agreements (in place of an implied threat of alternative government regulation) by developers that information on the thermal specifications of buildings be included in sales literature would have had potential if it had not already been overtaken by the EU Energy Performance Directive. Finally, it is worth noting that institutional development, while not a policy instrument as such, is very important. Energy efficiency is usually the concern of a number of government departments. In order to mobilise the policy process, it is helpful if a focal point for energy-efficiency to be established. Sustainable Energy Ireland provides a key role in this regard.

With record exchequer returns and budget surpluses, the Irish Government is in a position to embark on new investment programmes, provided that the returns justify the costs. In addition, many private households likewise have sufficient disposable income and capacity to borrow funds to undertake new investments. However, a household energy-efficiency strategy, which is comprised of a mix of instruments, would be required to mobilise the market to achieve the potential for domestic energy efficiency. Any household energy strategy should firstly distinguish between those households who have sufficient income to finance retrofitting conservation measures and those who don't. The former households will have relatively low discount rates and have savings that can be diverted to undertaking such an investment. Many of the least energy-efficient houses are occupied by low-income families. Full cost grants are likely to be necessary if these houses are to be encouraged to capture the benefits of energy efficiency. Much of the benefit will accrue to these households in the form of increased comfort and lower morbidity and mortality as a result of warmer homes, a marked improvement in terms of quality of life. In 2003, an important element of an overall strategy was announced. Sustainable Energy Ireland launched a €7.62 million initiative (funded under the National Development Plan 2000-2006) to address Fuel Poverty. Under the Low Income Housing Programme, community-based organisations will be funded to install energy efficiency measures into homes occupied by low-income households. The type of work to be undertaken through the programme will include the installation of attic and cavity wall insulation, draught proofing, hot water cylinder jackets and energy saving light bulbs or CFLs. Householders will also receive detailed energy advice that will enable them to maximise the benefit of any measures installed. Another key aspect of the programme will be facilitating the development of a national network for the delivery of longer-term solutions.

Conclusion

In summary, this discussion has demonstrated how the rapid growth of the Irish economy over the 1990s has led to rapid growth in energy consumption. This has resulted in increased pressure on environmental endowments, in particular, via significant increases in greenhouse gases and other air emissions. In urban environments, there are also significant impacts on human health. While the greatest growth has occurred in the transport sector, improving energy efficiency in the residential sector is of importance not just in terms of environmental improvement but also for reducing fuel poverty. The primary objective of this paper has been to examine the environmental implications of energy use in these sectors, to examine the factors which inhibit improved energy performance, and to provide some recommendations to improve performance. The results show that there are opportunities to improve performance in the transport sector at the micro level via parking pricing and possibly road pricing. However, the effectiveness of such instruments will depend on increasing price elasticities by addressing supply-side issues, in particular, the effectiveness and attractiveness of the public transport system. Spatial planning has an important role to play in addressing the relationship between spatial patterns of development, commuting patterns and energy use. However, the effectiveness of the National Spatial Strategy in promoting sustainable development now seems in doubt.

In relation to the residential sector and building performance generally, there is a need to make progress on improving the energy efficiency of the housing stock. Particularly important initiatives in this area include the EU Energy Performance Directive and the SEI fuel poverty alleviation programme and information campaigns. A major weakness in relation to macro-level environmental policy is that the EU Emissions Trading Scheme does not directly include the residential and transport sectors and the recent decision by the Irish government to abandon plans for a carbon tax which would have been targeted at small pollution sources (such as households) and mobile sources (such as vehicles) means that significant opportunities for reducing energy use and greenhouse gases are being lost.

Acknowledgements

I very much appreciate the helpful comments of the editors, Mark Scott and Niamh Moore and the referees. I am grateful to Frank Convery, John Healy, Andrew Kelly, and Eoin McLoughlin for allowing me to use data and results from previous joint work.

References

Arny, M., Clemmer, S. and Olson, S. (1998) *The Economic and Greenhouse Gas Emission Impacts of Electric Energy Efficiency Investments: Report 4 of the Wisconsin Greenhouse Gas Emission Reduction Cost Study*, Wisconsin: The Consortium for Integrated Resource Planning / University of Wisconsin / Leonardo Academy Inc. for the US Department of Energy and Oak Ridge National Laboratory.

Blasnik, M. (1998) *Impact Evaluation of Ohio's Home Weatherization Assistance Program: 1994 Program Year*, Ohio: Proctor Engineering Group.

Brechling, V. and Smith, S. (1994) 'Household Energy Efficiency in the UK', *Fiscal Studies*, Vol. 15(2), pp. 44-56.

Clinch, J.P. (2002). 'Reconciling Rapid Economic Growth and Environmental Sustainability in Ireland, the Barrington Prize Lecture', *Journal of the Statistical and Social Inquiry Society of Ireland*, Vol. XXX, pp. 159-218.

Clinch, J.P. and Healy, J. D. (1999) 'Alleviating Fuel Poverty in Ireland: A Programme for the 21st Century', *International Journal for Housing Science*, Vol. 23(4), pp. 203-215.

Clinch, J.P. and Healy, J. D. (2001) 'Cost-Benefit Analysis of Domestic Energy Efficiency', *Energy Policy*, Vol. 29(2), pp. 113-124.

Clinch, J.P. and Healy, J.D. (2000) 'Housing standards and excess winter mortality', *Journal of Epidemiology & Community Health*, Vol. 54(9), pp. 719-20.

Clinch, J.P. and Kelly, J.A. (2002). 'Traffic Congestion in Dublin: Past, Present and Future', in Convery, F. and Feehan, J. (eds) *Achievement and Challenge: Rio+10 and Ireland*, Dublin: University College Dublin, pp. 196-206.

Clinch, J.P., Convery, F.J. and Walsh, B.M. (2002). *After the Celtic Tiger: Challenges Ahead*, Dublin: O'Brien Press.

Collins, K.J. (1986) 'Low indoor temperatures and morbidity in the elderly', *Age and Ageing*, Vol. 15, pp 212-220.

Consultation Group on Greenhouse Gas Emissions Trading (2000) *Draft Report of the Consultation Group on Greenhouse Gas Emissions Trading*, Unpublished report.

Convery, F. (2001) *Movement and Congestion - What are the Issues?* Presented at 16th annual FFS conference Movement and Congestion: Transport Options for Ireland, Dublin: May 2001.

Curwen. M. (1991) 'Excess winter mortality: a British phenomenon?', *Health Trends*, Vol. 22, pp. 169-75.

DTO (2001) *A Platform for Change*, Dublin: Dublin Transportation Office.

EPA (2004) *Ireland's Environment 2004 - the State of the Environment*, Dublin: Environmental Protection Agency.

Evans, J., Hyndman, S., Stewart-Brown, S., Smith, D. and Petersen, S. (2000) 'An epidemiological study of the relative importance of damp housing in relation to adult health', *Journal of Epidemiology and Community Health*, Vol. 43, pp. 677-686.

Hardin, G. (1968) 'The Tragedy of the Commons', *Science*, Vol. 162.

Healy, J. (2001) 'Energy in Ireland: Economics and Policy Issues', *Environmental Studies Research Series working paper ESRS 01/10*, Dublin: University College Dublin.

Healy, J.D. and Clinch, J.P. (2004) 'Quantifying the Severity of Fuel Poverty, Its relationship with Poor Housing and Reasons for Non-investment in Energy-saving Measures', *Energy Policy*, Vol. 32(2), pp. 207-220.

Henderson, G. and Shorrock, L. (1989) *Energy Use in Buildings and Carbon Dioxide Emissions*, Watford: Building Research Establishment.

McLoughlin, E. (2001) *An Economic Analysis of Air Pollution Regulation: A Case Study of the Ban on Bituminous Coal in Dublin*, Unpublished MSc (Environmental Policy) thesis, University College Dublin.

Morgenroth, E. (2001) *Analysis of the Economic Employment and Social Profile of the Greater Dublin Region*, Dublin: Economic and Social Research Institute.

Pope C.A., Thun, M.J., Namboodir, M.M., Dockery, D.W., Evans, J.S., Speizer, F.E., Heath, C.W. (1995) 'Particulate air pollution as a predictor of mortality in a prospective study of US adults', *American Journal of Respiratory Critical Care Medicine*, Vol. 151, pp. 669-674.

Pezzey, J. (1984) *An Economic Assessment of Some Energy Conservation Measures in Housing and Other Buildings*, Watford: Building Research Establishment.

Ryan, L. (2004) *Strategy to Reduce Greenhouse Gases from Irish Transportation*, Dublin: Sustainable Energy Ireland.

Salvage, A.V. (1992) *Energy Wise? Elderly People and Domestic Energy Efficiency,* London: Age Concern.

Sckumatz, L. A. (1996) *Recognising All Programme Benefits: Estimates of Non-Energy Benefits from the Customer Perspective,* Washington: Skumatz Economic Research Associates Inc.

Stapleton, L., Lehane, M. and Toner, P. (2000) *Ireland's Environment: a Millennium Report*, Wexford: Environmental Protection Agency.

van Harmelen, A. K. and Uyterlinder, M. A. (1999) *Integrated Evaluation of Energy Conservation Options and Instruments,* Petten: Netherlands Energy Research Foundation.

Weber, G. (1990) *Earnings-Related Borrowing Restrictions: Empirical Evidence from a Pseudo Panel for the UK*, Department of Economics Discussion Paper 90-17, University College London.

Whyley, C. and Callender, C. (1997) *Fuel Poverty in Europe: Evidence from the European Household Panel Survey*, London: Policy Studies Institute.

Chapter 7

From Barricades to Back Gardens: Cross-Border Urban Expansion from the City of Derry into Co. Donegal

Chris Paris

Introduction: The Border Region - Poverty and Conflict to Building Boom

This chapter reviews some dramatic changes in the border zone between the City of Derry (subsequently 'Derry'[1]) and Co. Donegal since the mid-1990s: the border once had once been marked out with barricades but it is now demarcated by back-gardens in new housing estates in Co. Donegal.

The chapter builds on a pilot study of housing in the border counties (Paris and Robson, 2001), adding analysis of recent census and other housing data and introducing follow-up interviews with planning officials in summer 2003. The pilot study had been stimulated by the surge of housing construction after the demilitarisation of the border following the cease-fires announced in 1994 by the PIRA and loyalist paramilitaries. Border checkpoints were dismantled, sometimes seemingly overnight. Bridges were rebuilt across rivers and streams to reconnect communities that had been cut off from each other by demolitions, concrete blocks and barbed wire barricades.

The border region had long been characterised as a depressed area and the site of armed conflict (Cook et al., 1997; Robb, 1984) and the Irish National Development Plan (Government of Ireland, 1999) identified the border region as an area of high need. The militarised border had added to economic peripherality, with heavily fortified checkpoints on main roads and minor roads rendered impassable. O'Dowd et al. (1995, p. 274) argued that the Irish border area was 'the most violently contested border region in western Europe' and concluded that cross-border divisions had intensified rather than lessened in recent years (O'Dowd et al., 1995, p. 283).

In contrast to those long bleak years of economic decline associated with conflict and the militarised border, Paris and Robson (2001) identified many signs of substantial cross-border investment from Derry-Donegal in the northwest to the Dublin-Belfast corridor. One of the most striking of these changes was the way that Derry was overflowing into Co. Donegal. The military checkpoints between Derry and Donegal had all gone and new industrial enterprises were being established. Most noticeable of all, however, was the rampant new housing development.

Section 2 introduces the case study and considers some of its distinctive features. Section 3 reviews aspects of the wider context: metropolitan deconcentration and changing housing systems in Ireland. Section 4 focuses on the case study, reviewing qualitative research on changing housing developments in the Derry-Donegal border zone and census data on population and housing change. Section 5 concludes with some thoughts on directions of change.

The Derry-Donegal Case: Distinctive Features of the Case Study

In some ways the Derry-Donegal case of cross-border urban expansion is unique, as it is the only part of the border region where a city in Northern Ireland has sprawled across the border into the Republic. As with all case studies, moreover, Derry has distinctive features, especially the local ethno-religious geography and the extent to which its twentieth century urban expansion was constrained by a militarised border.

Derry's population characteristics have changed enormously since partition. It was included within Northern Ireland as an iconic Protestant citadel in the North West but over time, and especially during the recent Troubles, it increasingly became a Catholic city, as Protestants fled the west bank 'Cityside' of the River Foyle to settle in the east bank 'Waterside' suburbs or leave the area altogether. Catholics comprised 69.6 per cent of the district council population in 1991 compared to a Northern Ireland average of 38.4 per cent. Figure 7.1 shows that by the time of the 1991 census the River Foyle had become the local equivalent of Belfast's euphemistically named 'peacelines', separating the two 'traditions' within Northern Ireland.[2]

The population of the more densely built up urban area to the west of the River Foyle, constrained to the west by the border with the Republic, was overwhelmingly Catholic, with the exception of the small 'Fountain' enclave near the Protestant cathedral. There was a more 'mixed' population on the Waterside, albeit with high degree of local level ethno-religious residential segregation. The population of the City of Derry district council area became slightly more Catholic during the 1990s, from under 70 per cent in 1991 to 71 per cent in 2001,[3] compared with a Northern Ireland average of 43 per cent. The proportion of Catholics on the Cityside by 2001, however, was over 95 per cent. Only Newry and Mourne district council had a higher proportion of Catholics in 2001 (76 per cent).

The population of Derry's district council area grew strongly after 1980 with a recorded population increase of 7 per cent between 1981 and 1991, compared to a Northern Ireland average of 3 per cent. Derry's population grew considerably faster than the Northern Ireland average of 6.8 per cent between 1991 and 2001, with an overall increase of 10.2 per cent in the district council area. The number

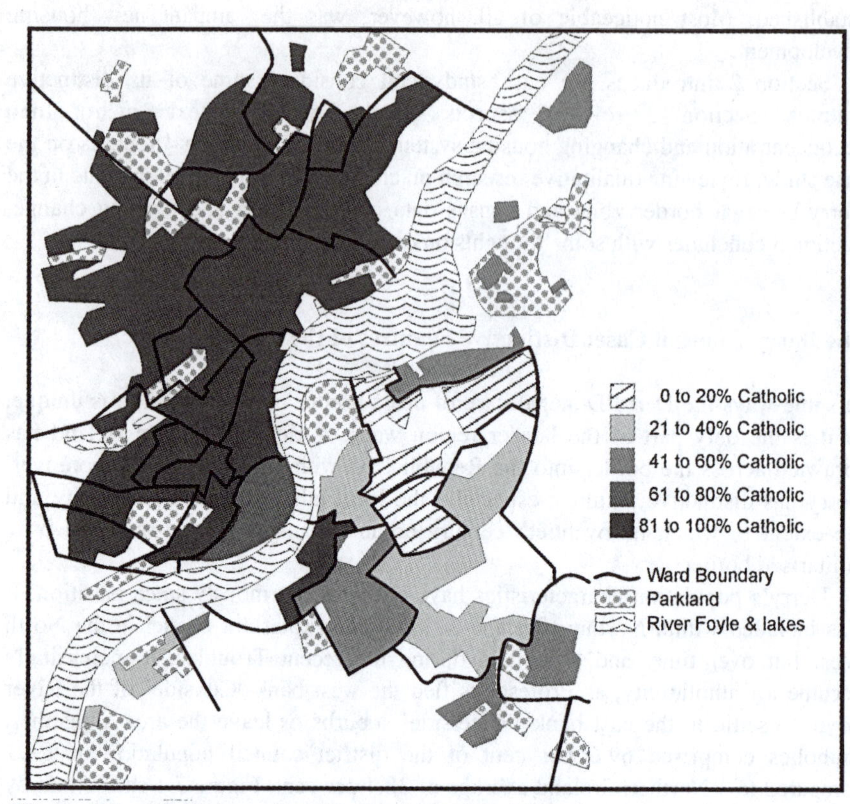

Figure 7.1 Religious distribution in Derry City

Table 7.1 Population change in Derry, 1991-2001

	Total population increase		*Roman Catholics*		*Other Christian denominations*		*Other religions, none & not stated*	
	Actual	**%**	**Actual**	**%**	**Actual**	**%**	**Actual**	**%**
Cityside	5,326	10.2	5,059	10.1	-681	-26.3	931	26.9
Waterside	4,377	9.5	3,164	19.3	-30	-0.2	1,243	39.8
Total	9,703	11.1	8,223	12.4	-711	-3.2	2,174	33

Source: 1991 and 2001 censuses

of households within the district council area, however, increased much faster than population: by over 28 per cent between 1991 and 2001.

The bulk of Derry's population and household growth was in the Catholic population as the number of persons overall stating membership of other Christian denominations actually fell by over 700 between 1991 and (see Table 7.1[4]). Table 3.1 also shows that the proportion of persons stating religion as 'none' or not stating any religion increased most strongly of all. The proportion of Catholics living on the Waterside increased substantially: the number of Catholics increased by 19 per cent whilst the number of those stating other Christian denominations fell marginally.

Above-average population and household growth were common in other mainly-Catholic parts of Northern Ireland during the 1990s, especially west of the Bann and in the southern border districts (Paris et al., 2004), but the second distinctive feature of the Derry-Donegal case is the city's proximity to the border and limited scope for expansion to the west other than by crossing the border.

The partition border had been drawn well away from the built-up area, to include the borough of Londonderry and its water catchment area as the only 'Northern' enclave to the west of the River Foyle. The city was physically constrained to the west by high ground from Creggan Hill to Sheriff's Mountain. Subsequent expansion to the northwest along the low ground either side of the A2 Buncrana road and north along the western shore of Lough Foyle resulted in Derry's western and north eastern suburbs approaching within a few hundred yards of the border, separated only by a narrow strip of so-called 'greenbelt' land. While the border remained heavily fortified, therefore, the only options for additional housing development lay on the other side of the River Foyle, or through densification of the existing built-up area.

No other city or large town in Northern Ireland is so close to the border nor have any been constrained by the existence of a fortified border. Enniskillen, with nothing like Derry's population growth, and with relatively easily developed land in all directions, is 15 km from the border. Armagh is also 15 km from the border with another 8 km to freestanding Monaghan. The nearest equivalent is Newry, but its southern suburbs are over 5 km from the border and there are no other major physical constraints to expansion.

This section has examined how the Derry-Donegal case study is distinctive, especially the distinctive ethno-religious geography of the city, with a rapidly increasing Catholic population highly concentrated on the west bank of the Foyle and – before the cease-fires – hemmed in to the north and west by a fortified border. In other ways, however, recent cross-border expansion of Derry is entirely consistent with broader changes that have been occurring within towns and cities on the island: these are examined in the next section.

The Context of Irish Urban, Regional and Housing Change in the 1990s

Irish cities and regions were transformed during the 1990s by physical restructuring and reorganisation of tenure systems (Paris, 2003). Nearly all towns and cities exhibited hollowing out of inner urban areas and rapid outward expansion comprising one-off homes in the countryside and low-density new private suburban development in 'commuter' locations (Department of the Environment for Northern Ireland, 1998; Horner, 1999; Government of Ireland, 2002). During the same period, most new housing was provided by the private sector, as social housing sectors declined

Metropolitan De-concentration

Paris and Robson (2001) characterised urban development in both jurisdictions of Ireland during the 1990s as metropolitan de-concentration. This idea is similar to the notion of 'counter-urbanisation' (see, for example, Breheny 1995; Champion et al 1998) but is favoured here because it identifies low density urban spreading as an emergent urban form not a movement towards non-urban living.

Census data for Northern Ireland (2001) and the Republic (2002) enable exploration of the extent of metropolitan de-concentration. The postponement of the Republic's census means that data for the two jurisdictions do not cover precisely the same time period. Even so, census data are satisfactory for analysis of overall trends, broad orders of magnitude and North-South comparisons. Table 2 contains an ad hoc regionalisation of the island of Ireland, partly following Horner (1993), but adapted in line with the availability of published data in the two jurisdictions and using census spatial categories. The regions as much as possible comprise units of approximately the same population size/share, mutadis mutandis the different scale of the two jurisdictions and the existence of more second tier cities within the Republic.

We can identify three broad zones for each of the two dominant metropolitan regions of Dublin and Belfast. The metropolitan cores comprising the cities of Dublin and Belfast respectively contained 14 per cent and 18 per cent of the population of their jurisdiction in 1991. Inner metropolitan zones each had around 15 per cent of the population of the jurisdiction. Outer metropolitan regions comprised 34 per cent of the population of Northern Ireland and, separated into Mid East and 'rest of Leinster', around 21 per cent of the population of the Republic. This regionalisation places Co. Louth within the border region, to facilitate cross-border comparative analysis.

Given the small overall size of Northern Ireland, much of its area can be considered as Belfast's commuting zone. In contrast, a much larger share of the Republic's territory is outside Dublin's commuter belt, either in the less densely populated far west, south and northwest, or within the ambit of Horner's (1993) 'embryo city-regions'. The fourth regional category comprises these secondary urban regions: Derry in Northern Ireland and the quartet of Cork, Limerick, Galway and Waterford in the Republic. For convenience, this analysis uses data aggregated at the district council level for Derry and combines each of the four Irish cities with their eponymous county.[5] The four Irish city regions comprised around 24 per cent of the national population in 1991 whereas Derry only accounted for 6 per cent of the Northern Ireland population, highlighting Belfast's dominance within the Northern Ireland space economy.

The other regional categories are 'border' zones in the two jurisdictions and the 'remainder', comprising the north coast in Northern Ireland and mainly midlands and western counties of the Republic. The border region was defined for convenience in the first cut analysis in terms of district council aggregates in Northern Ireland and counties in the Republic: accounting for about 10 per cent of the population of the Republic in 1991 but 20 per cent in Northern Ireland.

Table 7.2 shows regional population change in the two jurisdictions. Overall population increased by about 7 per cent in Northern Ireland and 11 per cent in the Republic. The availability of a 1996 census in the Republic, unlike Northern Ireland, reveals that population growth increased most strongly after 1996: from under 3 per cent between 1991 and 1996 to 8 per cent between 1996 and 2002. Official estimates of annual population change suggest that there was no equivalent sharp increase in Northern Ireland.

Despite different overall rates of population growth, data in Table 7.2 show that regional population trends in the two jurisdictions were remarkably similar during the 1990s and confirm the metropolitan de-concentration thesis. The population of inner metropolitan areas was static or falling. The net population change in the combined city and inner metropolitan zones of Dublin and Belfast represented a net declining share of about 1 per cent of the jurisdiction's population. In contrast, net population growth was very strong in outer metropolitan regions, especially west of Dublin where the Mid East region grew overall by 27 per cent between 1991 and 2002. Counties Kildare and Meath had population growth over 21 per cent between 1996 and 2002 alone. Strong growth was also evident in other regional cities, especially Galway and Derry (despite many of its families moving to Co. Donegal). Preliminary census analysis shows that, like Belfast and Dublin, the population of regional cities' central areas was generally static or falling, as population moved out to new suburbs or detached homes in outer areas. The population of Cork City, for example, fell by 3 per cent between 1996 and 2002 whereas Cork County grew by nearly 11 per cent. The 2002 census records similar processes of urban 'unpacking' at almost every spatial scale, even including small towns and villages (Paris, 2003). De-concentration was going on everywhere.

Changing Housing Systems

Housing provision in Northern Ireland and the Republic, by the early 1990s, had evolved in similar economic and demographic contexts (Paris, 2001). Both housing systems were characterised by low demand and slow house price growth, with little development pressure and widespread abandonment of rural dwellings, especially in the Republic. Slow economic growth and high unemployment in

Table 7.2 Regional population change in Northern Ireland and Republic of Ireland, 1991-2001/2

Region	***Census population (000)***		***Population change***		***Regional share of population change***
NI regions[1]	**1991**	**2001**	**Actual (000)**	**Per cent**	
Belfast City	279.2	277.4	-1.8	-0.6	-1.7
Inner Metro[2]	239.4	260.5	21.1	8.8	19.6
Outer Metro[3]	528.6	578.0	49.4	9.3	46.0
City of Derry	95.4	105.1	9.7	10.2	9.0
North Coast[4]	119.0	131.6	12.6	10.6	11.7
Other borders[5]	316.1	332.7	16.6	5.3	15.5
NI total	***1,577.8***	***1,685.3***	***107.4***	***6.8***	***100***
Republic regions	***1991***	***2002***			
Dublin City	478.4	495.1	16.7	3.5	4.3
Inner metro[6]	546.9	627.5	80.6	14.7	20.6
Mid East[7]	325.3	412.7	87.4	26.9	22.3
Rest of Leinster[8]	419.6	468.3	48.7	11.6	12.4
City regions[9]	844.3	934.1	89.9	10.6	23.0
Borders[10]	348.2	374.2	26.0	7.5	6.6
Rest of state	563.0	605.4	42.4	7.5	10.8
Rep. total	***3,525.7***	***3,917.3***	***391.6***	***11.1***	***100***

Notes:
1. Carrickfergus, Castlereagh, Newtownabbey and North Down
2. Antrim, Ards, Ballymena, Banbridge, Cookstown, Craigavon, Down, Larne, Lisburn & Magherafelt
3. Ballymoney, Coleraine, Limavady, Moyle
4. Armagh, Dungannon, Fermanagh, Newry & Mourne, Omagh and Strabane.
5. Source: Irish censuses 1991 & 20026. Dun Laoghaire-Rathdown, Fingal & South Dublin
7. Kildare, Meath, Wicklow
8. Excluding Co. Louth (now in Border RA) and Co. Sligo (counted within Rest of state)
9. Cork, Limerick Galway & Waterford (cities & counties)
10. Including County Sligo (formally now within Borders RA).

Source: Northern Ireland and Republic of Ireland censuses 1991 and 2001/2002

Northern Ireland resulted in the lowest UK house prices. Stop-start economic growth in the Republic during the 1960s and 1970s was followed by an economic crisis in the late 1980s and a return to population loss through net out-migration. The population of Northern Ireland had grown slowly, despite net out-migration.

Housing systems in the two jurisdictions had diverged since partition, though both had high levels of home ownership by EU standards. Home ownership was higher in the Republic in 1991, at nearly 80 per cent of households, with about 62 per cent in Northern Ireland. Housing policies had also differed, with stronger support for home ownership in the Republic and widespread council house sales for 50 years before Mrs Thatcher's right-to-buy legislation. Social housing provision differed markedly: one large statutory public housing authority, the Northern Ireland Housing Executive, accommodated around 28 per cent of households (down from 39 per cent in 1981, before the start of large-scale sales) whereas councils remained the public housing authorities in the Republic (accommodating around 10 per cent of households in 1991). Other social housing catered for about 2 per cent of households in Northern Ireland but under 0.5 per cent of households in the Republic. The private rented sector was of similar size in both jurisdictions, around 7-8 per cent of households.

Housing provision across the island was transformed during the 1990s. Private housing markets boomed and social housing provision declined in relative terms, especially in Northern Ireland. The 1990s building boom has been a noted feature of the Republic's Celtic Tiger economy (Clinch et al., 2002) although some commentators have remarked on increasing inequities within the Irish housing system (Drudy and Punch, 2002). Demand for new housing surged through a combination of economic growth, population growth and rapid growth of household numbers (Paris, 2003). Population growth in the Republic was fuelled by a dramatic shift from net out-migration to high levels of net in-migration after 1995. Natural increase and net in-migration between 1996 and 2002 together resulted in total population growth of nearly 300,000 in the Republic. In contrast, Northern Ireland mid-year population estimates indicated net migration loss of around 4,000 between 1991 and 2001, thus generating slower overall population growth than the Republic.

Growth in household numbers was much more rapid than population growth, especially in the Republic where households increased by over 26 per cent between 1991 and 2002. Even with slower population growth, Northern Ireland had household growth over 18 per cent. Surging growth in the household numbers, together with the booming economy, strong employment growth and falling levels of unemployment, especially in the Republic, resulted in massive increases in private house building in the 1990s. Private starts in the Republic grew from under 16,000 a year in 1988 to nearly 60,000 in 2002 and 2003.[6] Private housing starts in Northern Ireland grew at a slower rate, from around 6,000 to over 10,000 a year.

Despite greatly increased production, house prices grew strongly in the 1990s in the Republic. New and second hand home prices doubled between 1994 and

1998, and doubled again by 2002. The strongest growth was in Dublin but house prices increased throughout the country for all types of dwellings. House price increases were steady rather than spectacular in Northern Ireland, though the rate of increase accelerated after 1996, especially in Belfast. Many commentators relate Northern Ireland house price growth to the 'peace process' but its timing bears a remarkable similarity to the boom in the Republic.

The rate of social housing construction was much less than in the booming private housing sector, whilst there was continued sale of public sector dwellings, especially in Northern Ireland where sales greatly exceeded all new social housing construction. Overall social housing production fell in the north but grew steadily in the south after 1998. The net effect of the private building boom, the sale of public housing and weak growth of other social housing was to make Irish housing systems more similar to each other but also more 'private' during the 1990s. Table 7.3 shows the net effects of these changes between 1991 and 2001 (2002 in the Republic).

The social housing sectors lost share in both jurisdictions, from 10 to 7 per cent in the Republic and, albeit still at a higher level in Northern Ireland, from 32 to 21 per cent. Home ownership levels converged, with strong growth in Northern Ireland (from 62 to 70 per cent of households) but a fall in the Republic (80 per cent in 1991 to 77 per cent in 2002). The private rental sector share increased very strongly in the Republic between 1991 and 2002, from 7 to 11 per cent; it also increased in Northern Ireland, but only from about 6 to 7 per cent between 1991 and 2001. Some of the growth of private renting is investor-driven (Gray and Hillyard, 2003) but other extra supply may be generated by homeowners deciding to let out their former homes after moving to another place – including cross-border relocation in our case study.

Table 7.3 Household tenure in the Republic of Ireland (ROI) and Northern Ireland (NI), 1991-2001/2

	1991				**2001/2**[1]			
	Home-owners	***Private renting***	***Social renting***	***Other***	***Home-owners***	***Private renting***	***Social renting***	***Other***
ROI	80.2	7.0	9.7[2]	3.0	77.4	11.1	6.9	4.6
NI	62.3	5.8	32.0	1.0	69.6	6.7	21.2	2.5

Notes:

1. 2001 in Northern Ireland and 2002 in the Republic.

2. Local authority renting only in Republic of Ireland, NIHE and housing associations in Northern Ireland.

Source: 1991 & 2001 census in Northern Ireland; 1991 & 2002 census in the Republic

Metropolitan deconcentration and market-led tenure restructuring together generated a rapidly changing residential geography across the island, with some convergence overall in tenure patterns, and a scale of new private sector development that was wholly unprecedented and largely unexpected. Social housing hardly figured at all in the sprawling new housing developments around every town and city.

Derry Spills Over the Border into County Donegal

Paris and Robson's (2001) interviews with estate agents and public officials on both sides of the border during 1999 highlighted the widespread incidence of new housing development throughout the border region, especially in the Northwest, and indicated growing cross-border housing investment by households for their own use and by investors. Further interviews with planning officials in June 2003 revealed extremely high volumes of current housing development, and the growth in planning applications implied increasing demand and a continuing housing boom. Respondents in 1999 and 2003 saw few if any barriers to cross-border housing development.

Cheaper building land in Donegal at the start of the boom, and a more permissive planning system, fuelled the growth of Derry's suburbs into Co. Donegal. Furthermore, there was a remarkable coincidence between the timing of de-militarisation and a growing shortage of zoned building land within the City of Derry. Derry's rapid growth during the early 1990s had almost exhausted land supplies and planners were working on new development proposals to bring on the supply of additional building land. Just as there was growing pent-up demand from developers, anxious to cater for Derry's growing number of households, so the checkpoints came down and the roads were re-opened. New opportunities were seized by developers from both sides of the border, ranging from one-off detached homes along main roads, through small clusters of dwellings to substantial new suburbs in most border villages. Individuals and families, too, took advantage of a more relaxed planning regime and purchased building sites and contracted builders. The result has been the most dramatic building boom in ever in the southern Inishowen region of Co. Donegal. Even as some of the old shopkeepers were closing down for the last time, new suburban sub-divisions were being cut in Muff, Killea, Burnfoot and Bridgend. In February 2004 large new estates were coming onto the market in Newtowncunningham, half way between Derry and Letterkenny.

During the interview in 1999, Co. Donegal planners estimated that there had been an increase of 300 per cent in planning applications during the last few years (Paris and Robson, 2001). Estate agents in Derry saw Co. Donegal border villages as catchment areas for an enlarged Derry housing market. One Derry-based agent viewed the developments taking place on the Donegal side of the border as a 'logical extension of the city of Derry stretching into its natural hinterland'. He noted that many developers and builders had long records of activity on either side of the border. He also suggested that Derry had become attractive for outside speculative investors, mainly from the Republic, in particular those seeking to become involved in the rental market. Another estate agent estimated that 50 per cent of the total investment in some new projects came from investors based in the Republic.

Most respondents felt that the border was no longer the contentious issue that it had been before the 1997 all-party political Agreement. Planners and housing officials on both sides of the border were concerned primarily with the practical and concrete tasks of processing planning applications and managing the existing housing stock. Estate agents, however, emphasised the potential for development of the changing political and economic conditions of the border. In particular, in 1999 they argued that the strength of the pound against the punt made investment in Co. Donegal very attractive for homebuyers from the north. They also noted the absence of domestic rates in the Republic; their newspaper advertisements routinely emphasise the same point for prospective cross-border purchasers!

During interviews in summer 2003, we were struck by the continued intensity of the housing boom and the pressure on local planning officials in both jurisdictions. Local planners had limited knowledge of demographic trends or who were the main purchasers of newly constructed housing. Many of our questions – for example about in-migration, population trends, the extent of retirement in-migration, changing commuting patterns and house prices – were answered anecdotally or to the effect that 'we don't really know about that'.

But Co. Donegal planners did know which parts of their areas were experiencing most growth: Letterkenny, coastal areas and the border zone adjacent to Derry. As more zoned land had come on stream within the City of Derry, local planners there said that new housing was being developed 'everywhere' there was zoned land. Planners on both sides of the border agreed that it did not present any significant barrier to house builders and developers. Planners in a number of Republic border counties reported increased activity by 'Northern' developers and builders operating in their areas. In Derry, as well, planners noted movement in both directions, with builders from the Republic operating locally and local builders going across the border.

All planning officials interviewed considered that there would be strong continuing demands for housing development, that planning permission remained easier rather than harder to obtain, and that demand was unlikely to ease in the foreseeable future. They expected to see continued strong demand in the border area adjacent to Derry and they had no expectation of significant changes in the availability of land or volume of new building. Both planners and estate agents suggested that some households operate strategic 'addresses' in Derry to retain access to health, education and other services in Northern Ireland links (as well as cheaper car registration). It would be interesting to find out how – and where – they filled in their census forms and where they are registered to vote! This has been confirmed by local newspapers reporting complaints from Derry residents who resent their children being denied places in local schools when children from families who had moved over the border were given places.

Our interview evidence suggests that there has been considerable integration of housing markets across this part of the border. Many builders, developers and estate agents operate within both jurisdictions and building materials sell in what is effectively a single market. Nobody was aware of any significant difficulties involved in obtaining mortgage finance: most lending institutions based in Northern Ireland do not lend for purchase across the border, but they have associates, alliances or colleagues across the border that do so.

These changes are especially striking in new housing estates blossoming adjacent to the sites of former military checkpoints. Those stark icons of a heavily militarised border have been replaced by new homes on all of the roads out of the City Derry, especially towards Muff, Killea or Bridgend in Co. Donegal. As a visitor moves nearer to the border along those same roads, passing many disused petrol stations on the way, new housing can be seen in the border 'villages' just inside Co. Donegal. An article in a local newspaper captured the sense of the changing times:

> Twenty years ago border villages such as Bridgend, Burt, Burnfoot, Muff, Killea, Carrigans and St Johnston were dying on their feet. Shops were closing, petrol stations lying abandoned, emigration was rife (…) How that has changed.(…) there have been huge housing developments bringing life back to the villages.(…) the Elaghbeg Business Park in Bridgend, is testimony to the thriving local economy and is now home to a number of great homegrown businesses. Shops on the border have reopened and refurbished to look bigger and better than before for the deluge of customers.(…) Today the border villages are a hive of activity. Just visit at any time, night or day. Thousands of cars pass through them daily, many availing of the first class petrol stations on offer. And on Lotto night, just join the queue. City centre Dublin could hardly be busier. Right along the border, new businesses are opening, from hairdressers to warehousing. Even Muff has a Chinese restaurant – a sure sign of economic well-being. The feelgood factor has changed the social landscape totally. (Anon. 2003, p. 14).

Even allowing for the hyperbole - nightlife in Muff and Killea may occasionally be less lively than in central Dublin – the local newspaper was reporting real changes in the Derry-Donegal border zone, most especially on the Donegal side. The author's evidence may have been anecdotal and impressionistic, but the article nicely illustrated local awareness of the spectacular change since the de-militarisation of the border. As recently as five or six years ago, it was commonplace for motorists from the Republic to come to Northern Ireland for cheaper petrol, but sweeping changes in exchange rates and growing differentials in fuel tax have forced many Northern garages out of business while others have been re-invented as car-washes, fast food outlets and/or shops.

These expanding villages are functionally outer suburbs of Derry, with no connection to a 'rural' economy. The back gardens of many homes in these estates mark out the line of the border. Other new one-off detached homes are scattered along just about every road in the border zone beside Derry. These developments during the last five years have resulted in some of the fastest-growing electoral districts in the Republic of Ireland being located immediately across the border from Derry in Co. Donegal.

Table 7.4 shows net population change in Co. Donegal between 1996. Net population growth was significantly below the national average: 5.7 compared to 8 per cent. That growth was strongly concentrated in two areas. First, growth was heavily concentrated in new suburban developments growth in Letterkenny rural area (43 per cent of net county population increase between 1996 and 2002). A high location quotient of 4.9 for Letterkenny rural emphasises the concentration of net population growth. There was a second concentration, however, in Inishowen rural area, next to the border with Derry, with 28 per cent of net population growth and a location quotient of 1.4. Other parts of the county barely saw net population growth and many EDs actually lost population.

Table 7.5 shows that processes of deconcentration and suburbanisation identified at the national level were also operating at the local level within Inishowen, with slow growth in Buncrana urban and net loss in Carndonagh, but strong growth in Buncrana rural and Glentogher, near Carndonagh. Figure 7.2 also shows the strong population increases in EDs immediately adjacent to Derry or on main roads to Moville, Buncrana and Letterkenny: especially Kilderry (including Muff village), Burt, Birdstown, Three Trees and Killea.

Table 7.4 Population change in County Donegal, 1996-2002

	Actual	*Per cent*	*Share of growth*	*Location quotient*
Letterkenny urban	503	6.6	6.8	1.2
Letterkenny rural area	3,194	27.9	43.2	4.9
Inishowen rural area	2,086	8.0	28.2	1.4
Stranorlar rural area	988	4.7	13.4	0.8
Rest of county	618	1.0	8.2	0.2
Co Donegal	7,389	5.7	100	n.a.
State	291,249	8.0	n.a.	n.a.

Source: Census of Population, 2002

Table 7.5 Population change in Derry border zone, County Donegal, 1996-2002

	Actual	*Per cent*	*Share of growth*	*Location quotient*
Derry border zone				
Birdstown	133	20.7	6.4	3.6
Buncrana rural	432	22.3	20.7	3.9
Burt	245	26.0	11.7	4.6
Castleforward	89	10.8	4.3	1.9
Kilderry	401	41.6	19.2	7.3
Killea	143	13.6	6.9	2.4
Three trees	82	17.6	3.9	3.1
Inishowen rural	2,086	8.0		1.4

Source: Census of Population, 2002

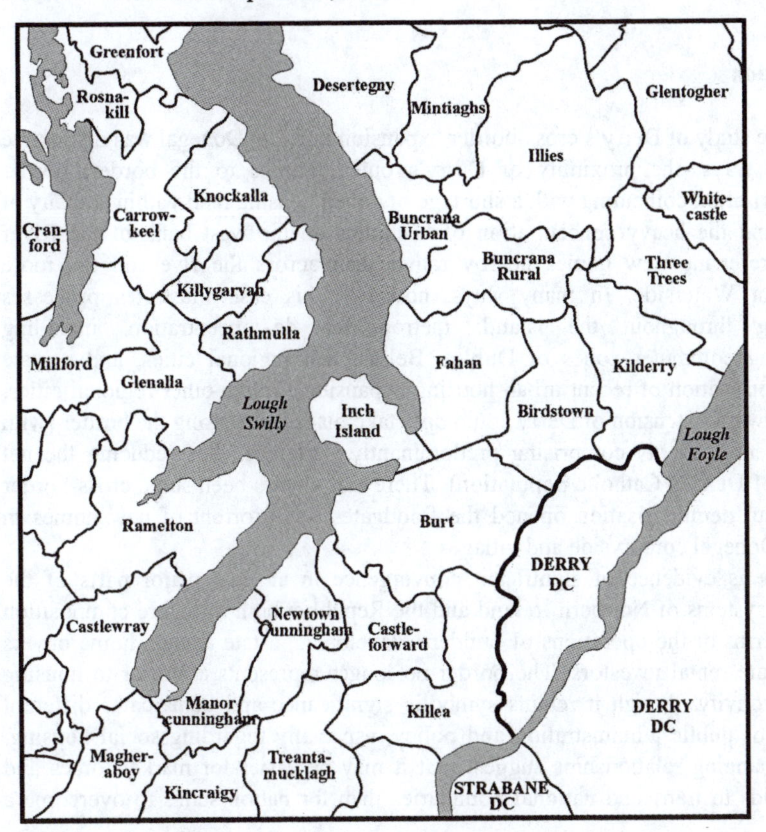

Figure 7.2 The Derry city-region

A casual drive through this area reveals a hive of construction activity and estate agents' advertisements in the Derry Journal testify to the continued development of new housing. The next census will undoubtedly see further large population increases in these and adjacent EDs. The 2002 census showed that there were population losses in many EDs in Inishowen; what does not show up in census data, however, is the explosive rash of holiday homes, which have been and are being constructed in these locations.

The development of Derry's Donegal suburban and ex-urban extension has been almost wholly a market-driven process. There has been very little new council housing development in this part of Donegal. Some parts of new housing estates have supposedly been 'affordable' housing, but prices have grown rapidly in line with general house price inflation. Recent interviews have even suggested that house price and land inflation in Co. Donegal may be leading to an inversion of the previous price differential, especially as large numbers of new houses are being constructed in Derry, and, especially with the Euro having appreciated by 10 per cent during the last year, these in turn may be priced below similar properties across the border!

Conclusion

This case study of Derry's cross-border expansion into Co. Donegal was distinctive in three ways: the proximity of Derry's outer suburbs to the border; border demilitarisation coinciding with a shortage of zoned housing land within the City of Derry; and the heavy concentration of Catholics on the west bank of the River Foyle preferring new homes nearby rather than across the river on the more Protestant Waterside. In many ways, however, this case illustrates processes operating throughout the island: metropolitan de-concentration, including expanding commuter zones of Dublin, Belfast and regional cities, and private sector domination of recent urban housing expansion. Unlike other regional cities, any westward extension of Derry could *only* take place by crossing the border (with the movers probably comprising predominantly Catholics, thus reducing the net growth of Derry's Catholic population). There had always been some cross-border living, but demilitarisation opened the floodgates for a torrent of new homes in nearby Donegal countryside and villages.

There is evidence of significant convergence in at least major parts of the housing systems of Northern Ireland and the Republic, both in tenure composition and in terms of the operations of builders, developers, estate agents, home buyers and private rental investors. The border no longer represents a barrier to housing market activity, though it retains symbolic significance and demarcates different systems of public administration and policy, especially regarding social housing. These changing relationships suggest that it may be easier for market forces and households to transcend national boundaries than for nation states to overcome a barrier which was created in fear and anger and which will remain in existence so long as those emotions dominate public policy in this small corner of the world.

Acknowledgements

I should like to thank my colleague Dr Terry Robson for his collaboration during the pilot study, in the re-working of findings of that study, and for undertaking the 2003 interviews of planning officials. This research has been supported by the Northern Ireland Housing Executive and the (then) Department of the Environment and Local Government and two University of Ulster Research Units of Assessment: Social Policy (1999-2000) and Build Environment (2003). None of these are responsible for any views expressed or errors contained in the present paper.

Notes

[1] The name of the city has been and remains contested. Many unionists insist that the correct name is 'Londonderry', although some of them use the term 'Derry' in casual conversation and all celebrate the feats of 'Apprentice boys of Derry' (emphasis added) during the eponymous siege by forces loyal to catholic King James. Nationalists primarily refer to 'Derry' although some will use the name 'Londonderry' from time to time to indicate goodwill to unionists. Sometimes the term 'Derry "stroke" Londonderry' (Derry/Londonderry) is used. This has become a trademark of a local media personality, Gerry Anderson, who often talks of 'Stroke City'. The official name of the district council, which includes both 'urban' and 'rural' areas is the City of Derry whereas the county remains 'Londonderry'. The Gaelic Athletic Association refers to the county as 'Derry' although there never was such a county either before or after the Plantation of Ulster, as counties did not exist as administrative entities in Gaelic Ireland.

[2] Extensive discussion of these processes of change, well illustrated by maps and photographs, can be obtained from the CAIN website at http://cain.ulster.ac.uk.

[3] Except where otherwise specified, all references to 1991 and 2001 data are from the Northern Ireland censuses of those dates and population data for the Republic derive primarily from the 1991, 1996 and 2002 censuses.

[4] Source: 1991 census (religion and gender by geographical area) and 2001 census Table UV018, both accessible on the NISRA website http://www.nisra.gov.uk.

[5] There are some boundary effects but they are trivial at the aggregate level.

[6] Source: data on housing completions derive from various editions of Housing Statistics from the Irish Department of the Environment, Heritage and Local Government and from Northern Ireland annual Housing Statistics from the Department of Social Development.

References

Anon (2003) 'Bordering on the brilliant', in the 'Inishowen Shopper' supplement to the *Derry Journal*, p. 14, 23 May.

Breheny, M. (1995) 'Counter urbanisation and sustainable urban forms', in Brotchie, J., Blakely, E., Hall, P. and Newton, P. (eds) *Cities in competition: productive and sustainable cities for the 21st century*, Melbourne: Longman, pp. 402-429.

Champion, T. (1989) *Counterurbanization: the changing pace and nature of population concentration*, London: Edward Arnold.

Clinch, P., Convery, F. and Walsh, B. (2002) *After the Celtic Tiger*, Dublin: O'Brien Press.

Cook, S., Poole, M. A., Pringle, D and Moore, A. (1997) *Deprivation in the Irish Border Region*, Belfast: Northern Ireland Voluntary Trust.

Department of the Environment for Northern Ireland (1998) *Shaping Our Future*, Belfast: Department of the Environment for Northern Ireland.

Drudy, P. J. and Punch, M. (2002) 'Housing models and inequality: perspectives on recent Irish experience', *Housing Studies*, Vol. 17(4), pp. 657-672.

Government of Ireland (1999) *Ireland National Development Plan 2000-2006*, Dublin: Stationery Office.

Government of Ireland (2002) *2002-2020 People, Places and Potential*, Dublin: Stationery Office.

Gray, P., Hillyard, P., McAnulty, U. and Cowan, D. (2002) *The Private Rented Sector in Northern Ireland*, Belfast: Northern Ireland Housing Executive.

Horner, A. (1993) 'Dividing Ireland into geographical regions', *Geographical Viewpoint*, Vol. 21, pp5-24.

Horner, A. (1999) 'Population dispersion and development in a changing city-region', in Killen, J & MacLaren, A. (eds) *Dublin: Contemporary Trends and Issues for the Twenty-First Century*, Dublin: Geographical Society of Ireland & Centre for Urban and Regional Studies, Trinity College.

O'Dowd, L., Corrigan, J. and Moore, T. (1995) 'Borders, national sovereignty and European integration: the British-Irish case', *International Journal of Urban and Regional Research,* Vol. 19(2), pp. 272-285.

Paris, C. (ed.) (2001) 'Housing in Northern Ireland, and comparisons with the Republic of Ireland', *Policy and Practice Series*, Coventry: Chartered Institute of Housing.

Paris, C. (2003) 'Housing markets and cross-border economic integration', Paper at the *Dublin-Belfast Corridor 2025* conference, Newry, September 2003.

Paris, C. and Robson, T. (2001) *Housing in the Border Counties*, Belfast: Northern Ireland Housing Executive.

Paris, C., Holmans, A. and Lloyd, K. (2004) *Demographic trends and future housing need in Northern Ireland, Final report to the Northern Ireland Housing Executive* (available from the Northern Ireland Housing Executive).

Robb, H. (1985) 'The border region: a case study', in D'Arcy, M. and Dickson, T. (eds) *Border Crossings: Developing Ireland's Border Economy*, Dublin: Gill and Macmillan.

Chapter 8

Urban-Generated Rural Housing and Evidence of Counterurbanisation in the Dublin City-Region

Menelaos Gkartzios and Mark Scott

Introduction

In recent decades, traditional patterns of urban growth and rural decline have changed dramatically in most advanced capitalist societies. In this transformation, the key factor has not been any important shift in the natural balance between births and deaths in either urban or rural areas, but rather it is the reversal from net loss to gain in the transfer of people from town to countryside that has been the crucial factor. The dominant net movement in advanced countries, Perry et al. (1986) argue, is now generally away from the older and larger towns and towards their rural fringes and the less densely populated outer regions. The purpose of this chapter is to explore in-migration trends to the rural hinterland of the Dublin city-region.[1] Recent trends have been characterised by population increases in Dublin's surrounding rural areas, resulting in urban sprawl and an increasingly dispersed city-region, which in turn fuels greater car dependency and longer commuting distances. This trend is not unique in Ireland, as similar spatial phenomena have been observed in most western societies and, in part, reflects changing demands for rural space (in an increasingly post-agricultural society) and consumer preferences towards rural areas and environmental amenities. However, in the case of Dublin, housing growth in the rural hinterland of the city not only includes the growth of rural towns and villages, but also of dispersed single dwellings in the open countryside – a distinctive feature of Irish settlement patterns. This chapter will begin with a discussion of the counterurbanisation concept, followed by a brief overview of counterurbanisation in a European context. The national and regional policy framework for managing settlement will then be reviewed. Drawing on recent census data, the chapter will then examine recent spatial trends in residential development in the Dublin city-region to consider the extent that residential growth has been occurring in the rural hinterland of Dublin City. This will involve four components: demographic changes; house-building activity; house prices; and travel-to-work commuting patterns. Finally conclusions are developed, in particular highlighting the need for further research.

The Counterurbanisation Phenomenon

The term counterurbanisation was first coined in North America, nearly thirty years ago to describe the spatial out-migration of people from cities to rural areas. For the American geographer Berry, counterurbanisation represented the direct antithesis of urbanisation. Counterurbanisation was defined as "a process of population deconcentration; it implies a movement from a state of more concentration to a state of less concentration" (Berry, 1976, p. 17). In contrast to the defining elements of 19th and early 20th century industrial urbanisation (increasing size, density and heterogeneity), the process of counterurbanisation has as its essence decreasing size, decreasing density, and a decreasing heterogeneity (Berry, 1976). The conceptualisation of the new term helped to draw attention to a new and important tendency, which constituted a turning point in the American urban experience.

The model of urban/rural movement was initially seen as a clean break from the past – a decisive break with traditional urban value-systems. It constituted the complete antithesis of conventional thought of urbanisation: centrifugal rather than centripetal, operated by quality of life instead of pecuniary forces and consumer rather than producer-led. The theory suggests that with higher incomes, greater increased mobility and a vastly improved rural infrastructure, large numbers of town-dwellers can move towards their ideal of a rural lifestyle. Advanced capitalist societies have entered a high-mobility era where masses of people can live and work where they choose. Thus, this kind of movement involves not merely a geographical but also an ideological leap, and a desire for an alternative way of life (Perry et al., 1986). Generally these changes fitted very closely with notions of a shift from an industrial to some form of 'post-industrial' society, as they appeared to provide a physical and readily measurable manifestation of more complex and deep-seated changes believed to be taking place in economic and social structures (Champion, 1998). The spillover school of thought represented the first and least restrictive step in narrowing down the definition of counterurbanisation from the very wide interpretation Berry gave. The central argument is that population growth taking place within the commuting fields of existing metropolitan centres cannot be considered as a break from past trends, but merely a continuation of the local metropolitan decentralisation involving shifts of people and jobs from core to ring in each urban region (Champion, 1989). According to the urban-sprawl model higher-paid and prestigious 'head-office' functions continue to concentrate in the centre while lesser-skilled operations are forced out because of rising rent and labour costs. Workers also move to the rural periphery, taking advantage of the cheaper housing cost and the advances of telecommunication, which allow 'net-workers' to communicate rather than to commute to work over long distances (Perry et al., 1986).

Most studies seeking to explain counterurbanisation focus on the motivations of people moving into rural areas (Champion, 1989; Halliday and Coombes, 1995; van Dam et al., 2002). However, there has been little precise agreement as to the significance of the term itself. Counterurbanisation has been interpreted either as a migration movement, a relocation of urban residents from large (often metropolitan) to small (often non-metropolitan) places or a process of settlement change. The focus, therefore, shifts from a migratory movement to one of a process of settlement system change, resulting in the creation of a deconcentrated settlement pattern (Mitchell, 2004). In an effort to promote a more consistent usage of the word, Mitchell construed counterurbanisation as a migration movement, rather than a pattern or a process.[2] Such a movement cannot be considered homogenous but may vary in terms of both the destination of migrant households and the motivations driving relocation – therefore its categorisation in sub-forms is necessary. The three forms Mitchell proposed are as follows:

- The term ex-urbanisation can be used to describe the movement of well-to-do urban dwellers in the bucolic countryside surrounding urban centres. Although motivated to reside outside the metropolitan core, these ex-urban residents are highly connected to city with daily commute to work;
- The term displaced-urbanisation can describe household moves that are motivated by the need for new employment, lower costs of living and/or available housing. Unlike ex-urbanisation, moves taken by such urban dwellers are to whatever geographic location provides for these needs;
- The term anti-urbanisation can describe movement of people whose driving force is experiencing life in a non-metropolitan setting. Anti-urban motivations move residents (beyond the suburbs) to escape crime, taxes, congestion and pollution. Therefore, in this case urban dwellers not only long to live in a rural environment (as a result of push and pull factors) but, for those in the labour force there is also the desire to work in a less concentrating setting.

Counterurbanisation in Europe

According to Kontuly (1998), spatial trends in European countries between the 1960s and the 1990s showed no clear-cut patterns but rather trends towards urbanisation as well as counterurbanisation. In Finland for instance, urban concentration is a dominant feature of the migration system; migration flows are heavily directed towards the few largest urban centres located mainly in the southern parts of the country. However, against the common belief that migrants head only to urban areas, both urban and rural regions are experiencing in and out-migration, and there is also a constant inflow of migrants to peripheral and more distant regions. In-migration to rural areas in Finland is selective, but partly in an atypical way as rural in-migrants tend to be older and have less human capital than those moving to other areas. Therefore, rural areas are constantly losing the most competent (young, educated) segment of their population to urban regions and instead they receive retirees. Despite the current trend of migration, opinion polls show that two out of three Finns place a premium on rural residential environment against the urban. In particular it is the residential preferences that are drawing people into rural areas: a good living environment, cheaper housing and quality of life are the strongest pull factors of rural areas (Nivalainen, 2003).

Similarly in the Netherlands the demand for rural residential environments appears to be large. Urban-rural migration is also a selective process in the Netherlands, and it involves higher income households, the elderly, and younger and middle-aged families with children. In addition, urban-rural migration is tightly constrained by factors such as the availability of dwellings, house prices and development control and regulation at a local level. In areas that are commonly perceived as attractive, rural housing is a scarce commodity and this scarcity is maintained by state planning and intervention. Most people who consider a move to rural areas, consciously and deliberately choose a rural environment on the basis of the 'rural' characteristics of the area. However, new possibilities for rural housing are currently restricted and consequently, Dutch spatial policy is facing the dilemma whether this demand for rural residential environments should be facilitated and how (van Dam et al., 2002).

In the UK evidence of previous censuses highlights a continuing decline in population in large urban areas and an increase in population in rural areas. These changes in population have been accompanied by shifts in employment and retailing but evidence suggests that the dispersal is associated with longer travel distances and increased reliance on private transport. The quality of the rural environment is one of the most important factors in the appeal of rural areas as places to live and many city dwellers aspire to live in more rural areas (Stead, 2002).

From this brief assessment of European experiences, it is apparent that counterurbanisation cannot be evaluated without considering the concept and perception of rurality. As van Dam et al. (2002) argue, urban-rural migration, images and representations of the rural and preferences for living in rural residential environments are strongly linked. The manner in how people conceive rurality – a rural idyll as many commentators argue (see for example, Halfacree and Boyle, 1998; Gorton et al., 1998; Boyle et al., 1998) – can influence individual migration behaviour. Thus, images of the countryside play a considerable role in rural living preferences and can make rural residential environments very popular amongst urban residents. However, images and representations of the rural are not static and have changed considerably as rural economies and societies continue to restructure. In fact, fundamental transformations have taken place in Europe's rural economy and society, and new patterns of diversity and differentiation are emerging within the contemporary countryside. These may be summarised as (drawing on Marsden, 1999):

- The decline in agricultural employment, and in the relative economic importance of food production, accompanied by structural changes in the farming industry and food chain;
- The emergence of environmentalism as a powerful ethic and political force;
- The related emergence of new uses for rural space, and new societal demands in relation to land and landscape and the treatment of animals and nature;
- Increased personal mobility, including commuting, migration, tourism and recreation;
- The emergence of new winners and losers from change processes, and especially recognition of 'excluded groups' suffering from poverty and economic and social vulnerability.

Given the depth and prolonged character of crisis in the agricultural sector, some commentators have suggested that rural areas are experiencing a shift from a 'productivist' (agricultural) to a 'post-productivist' era in the countryside (Halfacree, 1997; Hadjimichalis, 2003). In this post-productivist phase, rural localities are now places that people from outside come into to consume the diversity of things that now make and constitute rural space (Gray, 2000). Traditional rural images based on agricultural features, and negative associations with agricultural issues (overproduction, environmental pollution, intensive farming, livestock diseases, etc.) and the presumed conservative rural life, whether correct or not, seem to have been replaced by broadly shared positive images connected with rural amenities such as greenness, nature, peace and quite and space (van Dam et al., 2002).

In line with the European experience, Ireland's rural communities are also undergoing rapid and fundamental changes: the agricultural sector continues to restructure; the economic base of rural areas is diversifying; new consumer demands and practices have emerged; there is a growing concern for the environment and increased pressure to include the environmental dimension in decision-making; and some rural communities are under intense pressure from urbanisation, while other areas continue to experience population decline. Within this context, a survey of 1,500 respondents[3] on public attitudes to the environment and quality of life issues (UII, 2001), revealed a particular image of the Irish countryside. The principal benefits of rural living included the peace of rural areas (18 per cent of all respondents), followed by clean air (13 per cent) and then space. Other perceived benefits of rural life included less drugs, pleasant walks, integrated schools and good sporting facilities. On the other side the main limitations of rural life in Ireland, according to the same report were the lack of public transport (29 per cent of all respondents), followed by isolation (15 per cent) and distance from facilities (9 per cent). Interestingly, some 9 per cent also felt the there were no limitations to rural living at all. Finally over half of respondents expressed a preference for rural life, whereas the corresponding figure for urban/suburban life was just over one third (Figure 8.1). The research also highlighted respondents' aspirations to shift from urban/suburban living in both the short to medium term to a more rural based lifestyle in the long term (Table 8.1).

Following this survey it appears that there is a preference in Ireland towards rural living. That preference could translate into an Irish rural idyll – an image of the Irish countryside appealing to the majority of people as a desirable residential environment, which represents both a desire among rural dwellers to remain in rural communities, and for urban dwellers to migrate to rural areas.

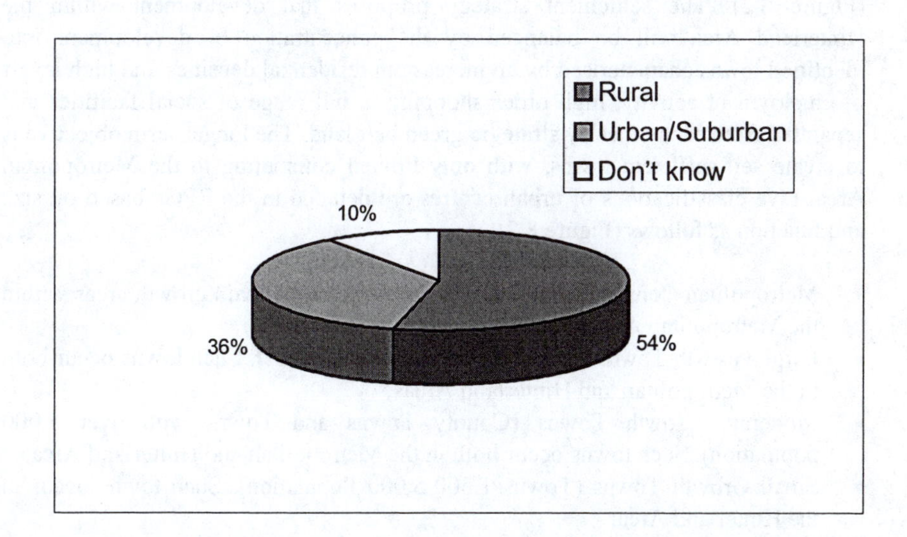

Figure 8.1 Preferences for rural or suburban lifestyle in Ireland

Table 8.1 Choice of areas in the short, medium and long term

Periods of time	*Choice of area (%)*		
	Rural	***Urban/Suburban***	***Don't know***
Next 5 years	39	57	4
5-10 years	38	52	10
More than 10 years	42	43	15

Source: *Public Attitudes to the Environment and other Issues* (UII, 2001)

Policy Framework in Ireland

Following European trends, the urban policy discourse in Ireland increasingly favours a compact city approach with increasing residential densities as a means to prevent urban sprawl and to reduce the spatial separation of daily activities, particularly travel-to-work patterns. Both the recently published National Spatial Strategy (NSS) (DoELG, 2002) and the Regional Planning Guidelines for the Greater Dublin Area (RPGs) (Dublin Regional Authority and Mid-East Regional Authority, 2004) endorse this approach to urban development. In relation to the Greater Dublin Area, the RPGs' settlement strategy proposes two separate Development Policy Areas – the Metropolitan Area and the Hinterland Area (Figure 8.2). The settlement strategy proposes that development within the Hinterland Area will be balanced by the concentration of development into identified towns characterised by an increase in residential densities and high levels of employment activity, high order shopping, a full range of social facilities and separated from each other by strategic green belt land. The longer-term objective is to create self-sufficient towns, with only limited commuting to the Metropolitan Area. Five classifications of urban centres are detailed in the RPGs based on size and function as follows (Figure 8.2):

- Metropolitan Consolidation Towns. They will be the main growth areas within the Metropolitan Area;
- Large Growth Towns (Primary Development Centres). Such towns occur both in the Metropolitan and Hinterland Areas;
- Moderate Growth Towns (County Towns and Towns with over 5,000 population). Such towns occur both in the Metropolitan and Hinterland Areas;
- Small Growth Towns (Towns 1,500-5,000 Population). Such towns occur in the Hinterland Area;
- Villages (Villages 1,000 Population): (a) Commuter Villages and (b) Key Villages. Such villages occur in the Hinterland Area.

However, despite this policy agenda, a distinctive feature of both remote and accessible rural areas in Ireland over the past thirty years has been the growth in dispersed housing in the countryside, and recent years have witnessed increasing difficulties in addressing this issue. The proliferation of dispersed single dwellings in the countryside has been an issue for many years, but the scale and pace of recent years appears to be intensifying. For example, analysis undertaken during the preparation of the National Spatial Strategy suggested that between 1996-1999 over one in three new houses built in the Republic of Ireland have been one-off housing in the open countryside, and highlighted that the issue of single applications for housing in rural areas has become a major concern for most local planning authorities (Spatial Planning Unit, 2001). Commentators such as Aalen (1997) and McGrath (1998) have argued that the planning system is unable to

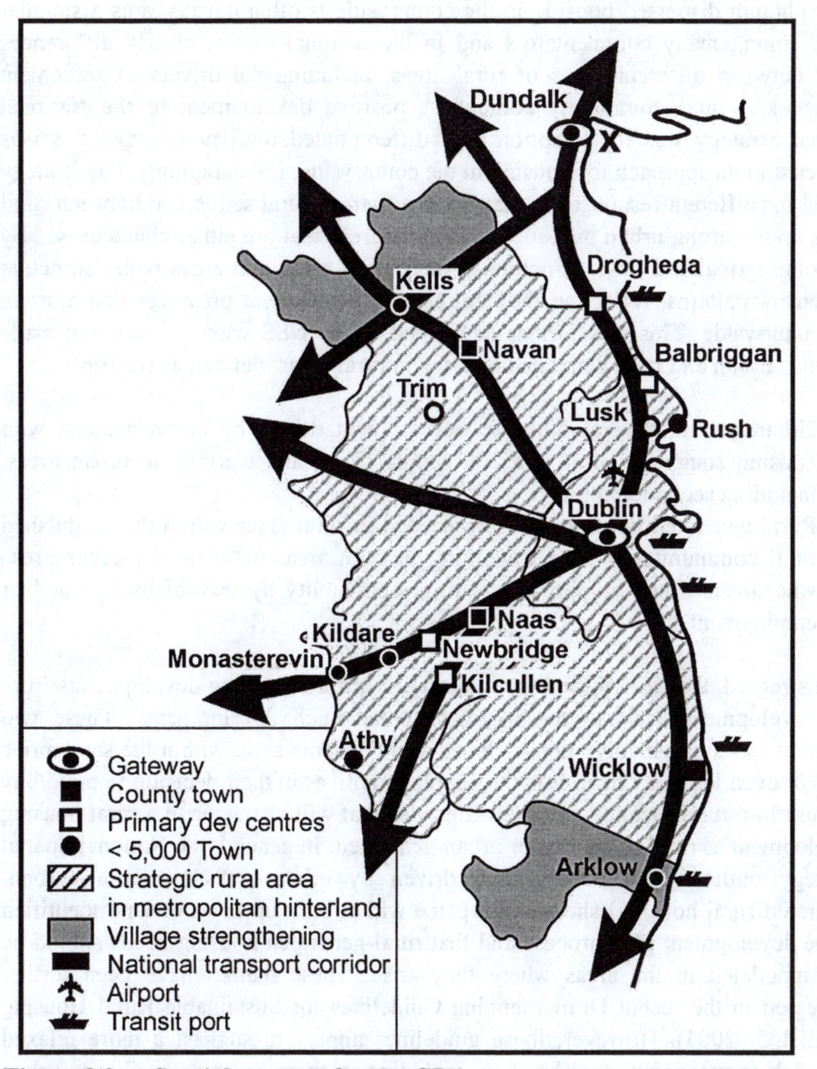

Figure 8.2 Spatial strategy for the GDA

respond effectively to rural settlement growth. In a critique of rural planning, both commentators suggested policy has been driven by the priorities of a few individuals, an intense localism, and the predominance of incremental decision-making. Similarly, Gallent et al. (2003, p. 90) classified rural planning in Ireland as a laissez-faire regime, suggesting that: "the tradition of a more relaxed approach to regulation, and what many see as the underperformance in planning is merely an expression of Irish attitudes towards government intervention."

Although dispersed housing in the countryside is often portrayed as a singular issue among many commentators and in the national media, clearly differences exist between different types of rural areas, including the drivers of settlement change and rural community context. A positive development in the National Spatial Strategy was the adoption of a differentiated rural policy, and this was reflected in its approach for housing in the countryside. Encouragingly, the Strategy called for different responses to managing dispersed rural settlement between rural areas under strong urban influences and rural areas that are either characterised by a strong agricultural base, structurally weak rural areas and areas with distinctive settlement patterns, reflecting the contrasting development pressures that exist in the countryside. This was further developed in the NSS with a distinction made between urban and rural generated housing in rural areas, defined as (p. 106):

- Urban-generated rural housing: development driven by urban centres, with housing sought in rural areas by people living and working in urban areas, including second homes;
- Rural-generated housing: housing needed in rural areas within the established rural community by people working in rural areas or in nearby urban areas who are an intrinsic part of the rural community by way of background or employment.

In this regard, the NSS shifted the importance of rural housing developments from the development itself to the motives behind such developments. These two different types can involve houses located in the same area, within the same price class or even look identical. It is the peoples' motives in their decision to build/buy a house in a rural location and their lifestyles that will characterise a rural housing development as rural-generated or urban-generated. In general, the National Spatial Strategy outlined that development driven by urban areas (including urban-generated rural housing) should take place within built up areas or land identified in the development plan process and that rural-generated housing needs should be accommodated in the areas where they arise. These themes have been further addressed in the recent Draft Planning Guidelines for Sustainable Rural Housing (DoEHLG, 2004). However, these guidelines appear to suggest a more relaxed approach to managing rural housing, including in those areas accessible to urban centres. In summary, the guidelines outline that (p. 1):

- People who are part of and contribute to the rural community will get planning permission in all rural areas, including those under strong urban-based pressures, subject to the normal rules in relation to good planning (related to site layout and design);
- Anyone wishing to build a house in rural areas suffering persistent and substantial population decline will be accommodated, subject to good planning practice in siting and design.

In relation to managing rural housing in the Dublin city-region, this emerging policy framework perhaps presents 'mixed-signals'. On the one hand, national and regional spatial strategies propose that residential development in the rural hinterland of Dublin should be concentrated in identified towns and service villages. However, on the other hand, the rural housing guidelines suggest a more permissive approach to dispersed housing in the countryside, including areas in close proximity to urban centres.

Residential Trends in the Dublin City-Region

This section will examine recent residential spatial trends in the Dublin city-region to consider the extent that residential growth has been occurring in the rural hinterland of Dublin City (Figure 8.3 for study area). This will involve four components: firstly, this section will outline recent demographic changes in the city-region and in individual local authority areas. Secondly, house building activity will be considered, focusing on private detached houses. Thirdly, house prices will be briefly discussed, and finally travel-to-work commuting patterns will be highlighted.

Demographic Changes

According to the last 5 censuses (after the late 1970s), it appears that the population has been growing constantly in all counties within the city-region until 1986. Whereas Kildare, Kilkenny, Meath, Wicklow and Dublin increased their population during 1986-1991, the remainder of counties experienced population decline during this period. Nevertheless after 1991 they all started growing – except the relatively remote Co. Longford. In particular counties Meath, Kildare and Wicklow started growing significantly faster than Co. Dublin and the national rate of growth after 1991. The trend continued after 1996 with Co. Dublin (including Fingal, South Dublin, Dublin City and Dun Laoghaire-Rathdown) growing more slowly (6 per cent) than the nation as a whole (8 per cent) and less rapidly than nearby counties within commuting distance to Dublin (Figure 8.4). For a comprehensive spatial analysis of the recent demographic changes in the city-region, population levels from the 2002 Census were analysed and read into Geographical Information System (GIS) maps. The smallest administrative area for which population levels were calculated and mapped was the Electoral Division (ED).[4] In Figure 8.5 percentage classes are applied to describe population changes from 1996 to 2002 in the Dublin city-region. Multiples of 8 have been used for easier comparisons of the growth of an area with the national average percentage growth (8 per cent). The most dramatic increases of population (3 times or more than the national rate of growth) have taken place in the mid eastern counties of the city-region – Counties Meath, Kildare, Offaly, Westmeath and Laois – and along the eastern coastline –

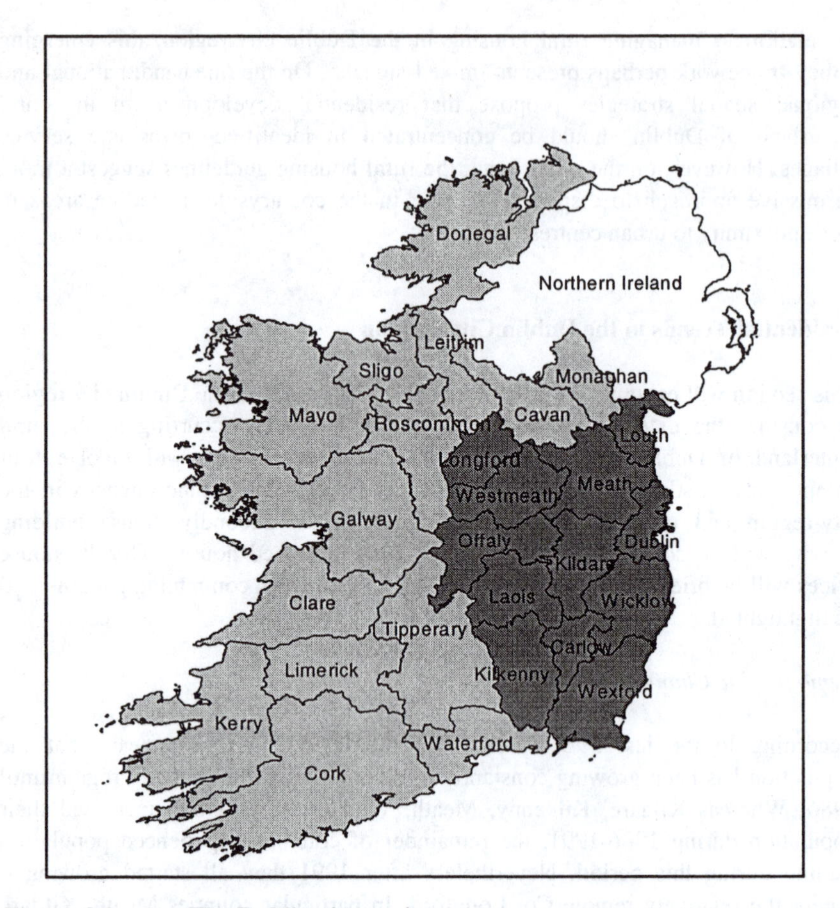

Figure 8.3 The Dublin city-region

Counties Louth, Wicklow and Wexford. Outside these counties, most areas in the city-region grew less than the national average or even decreased their population. The redistribution of population in the city-region suggests a residential preference for rural areas accessible by major road links and rail transport and a widening of the Dublin commuter belt well beyond the Greater Dublin Area, from Dundalk to Gorey and as far as inland as Athlone.

For a more detailed assessment, the maps below illustrate spatial changes for some of the counties that have experienced significant levels of population changes outside Co. Dublin (Louth, Meath, Kildare and Wicklow). Co. Louth (Figure 8.6) increased its population by 10.5 per cent from 1996 to 2002 with higher increases in the southern parts of the county, outside Drogheda. The spatial distribution of growth demonstrates that the rural environs of Dundalk and Drogheda experienced higher increases of population than their urban areas. For example Dundalk town increased its population by 4.5 per cent whereas the

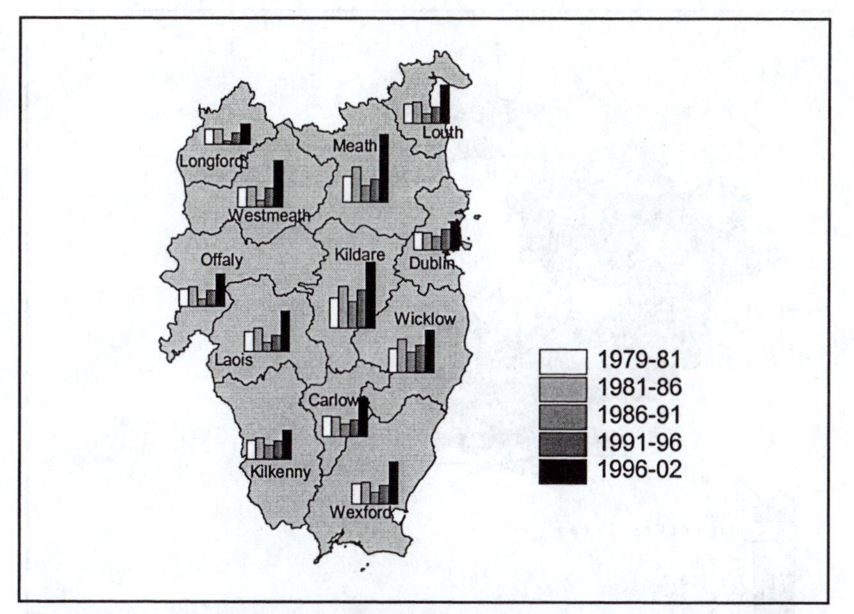

Figure 8.4 Percentage change in the population of each county in Leinster at each census since 1979

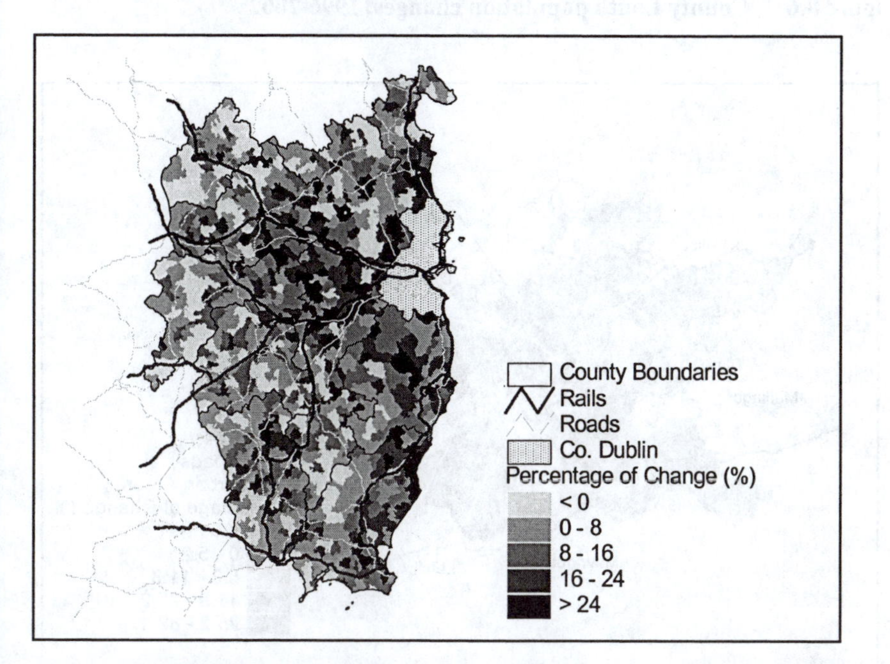

Figure 8.5 Percentage change in the population of electoral divisions in Leinster, 1996-2002

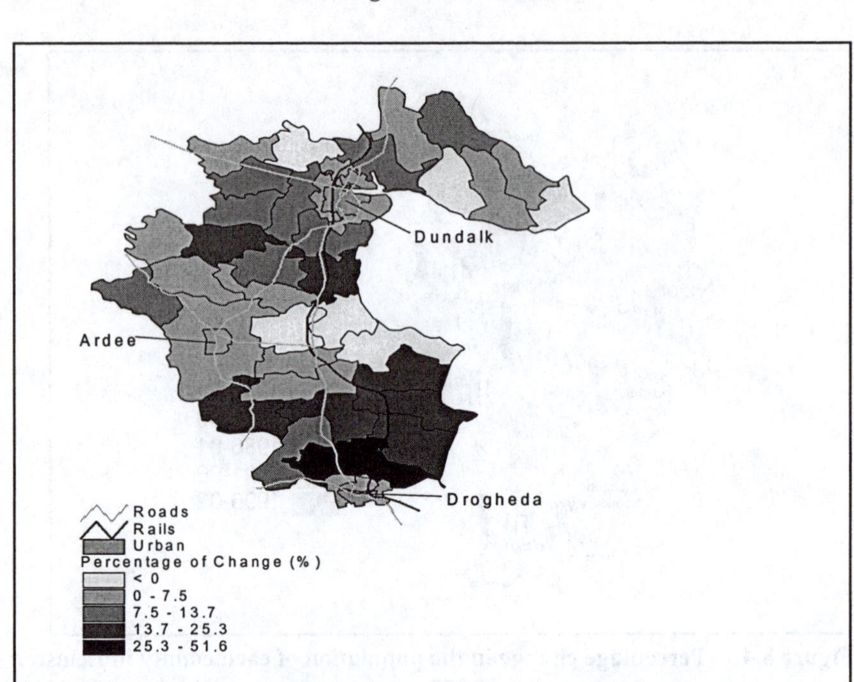

Figure 8.6 County Louth population changes, 1996-2002

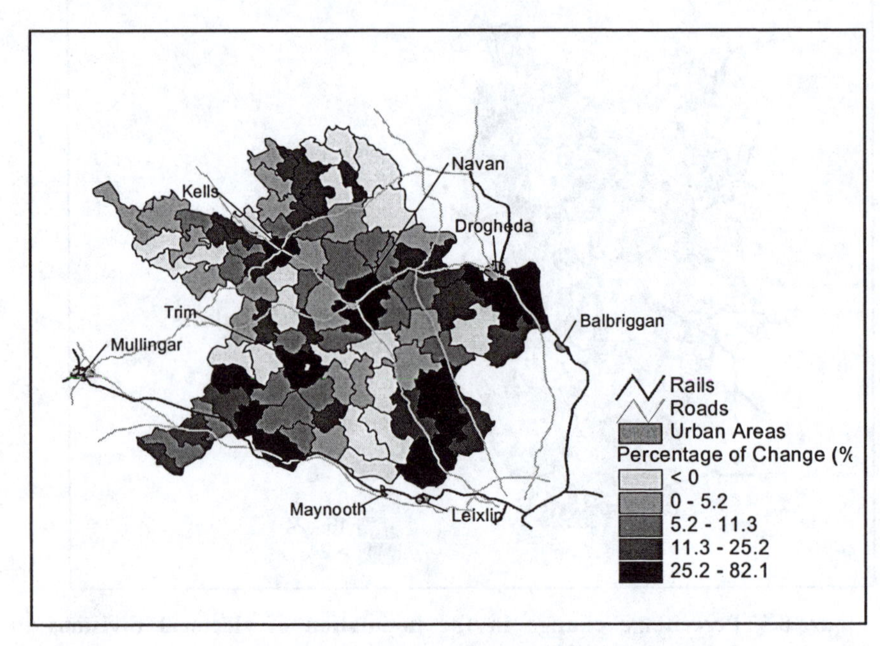

Figure 8.7 County Meath population changes, 1996-2002

adjacent rural areas increased two times more than the town itself. Similarly the urban area of Drogheda increased its population by 3.8 per cent, while some of the rural areas around Drogheda increased their population by almost 50 per cent. Co. Meath (Figure 8.7) increased its population more than any other county in the GDA, almost 3 times more than the national average (22.1 per cent). Major population increases took place close to the boundary with Fingal local authority area (Co. Dublin) and around towns like Drogheda, Navan, Kells and Trim. In all cases the population of these urban areas increased at a lower rate than their surroundings, except in the case of Trim where the town itself showed a significant decrease of population (-16.8 per cent), while its rural surroundings increased their population by 39.7 per cent. Despite the high levels of population increases in the county as a whole, 21 EDs (22.8 per cent of 92 EDs) decreased their population or remained stable, exhibiting a considerably unequal redistribution of population in Co. Meath away from small towns and villages and towards the rural environs of larger towns.

Similarly in Co. Kildare (Figure 8.8) the rural environs of Naas experienced higher increases of population than the town itself. Athy is the only urban area in the study area that increased its urban population rather than increases in its rural environs. The redistribution of population in Co. Kildare appears to be more even than in Co. Meath (only 7 EDs decreased their population). EDs that decreased their population appear to be the least well-served areas along the road and rail transport. In Co. Wicklow (Figure 8.9), population increases occurred near and around urban areas like Wicklow, Arklow and Bray. Co. Wicklow increased its population less than the other counties adjacent to Co. Dublin, possibly as a result of stricter planning policy related to its outstanding landscape quality.

Outside the GDA, major population increases follow the same pattern of residential development, along the rail and road transport links and near larger towns. For instance, in Co. Westmeath the rural environs of Mullingar and Athlone experienced much higher increases of population than their urban areas. Dramatic changes have taken place also in the periphery of Portlaoise (Co. Laois), which increased its population by 41.3 per cent, while the town itself recorded a marginal population decrease (-1.4 per cent).

House-Building Activity

Apart from population levels, it is also useful to assess housing activity in the study area. Figure 8.10 shows private house completions in the city-region (excluding Local Authority and voluntary/non profit houses). Private house completions tended to increase throughout the city-region. Co. Dublin, as expected, has had the highest housing completions followed by the adjacent counties and then the outer counties. Outside of Co. Dublin, Co. Meath was the strongest residential target in the GDA in 2003, followed by Co. Kildare (with 3,519 and 2,824 completions respectively). Significantly, Co. Meath increased its private house completions by 163 per cent in only 5 years (1999-2003).

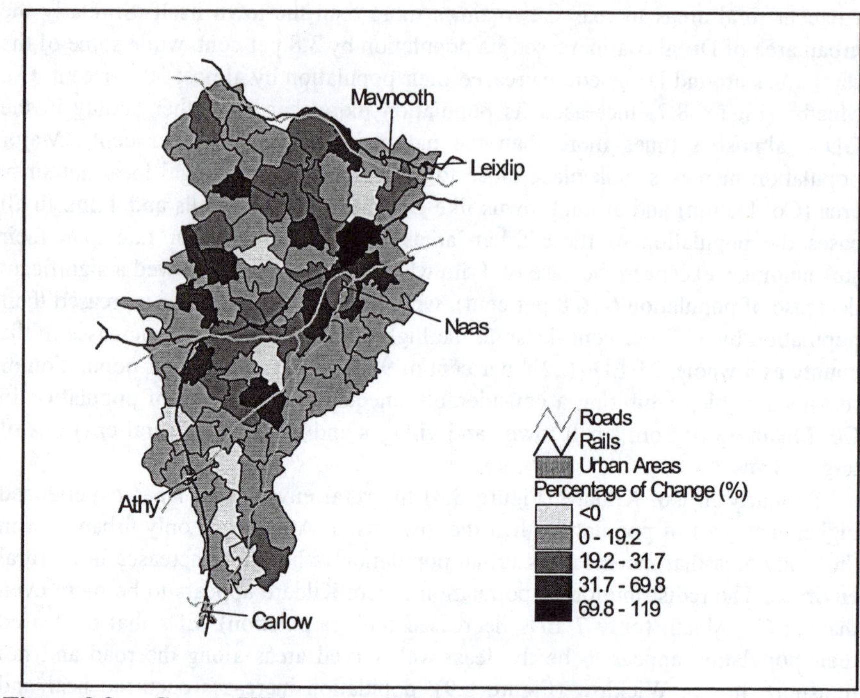

Figure 8.8 County Kildare population changes, 1996-2002

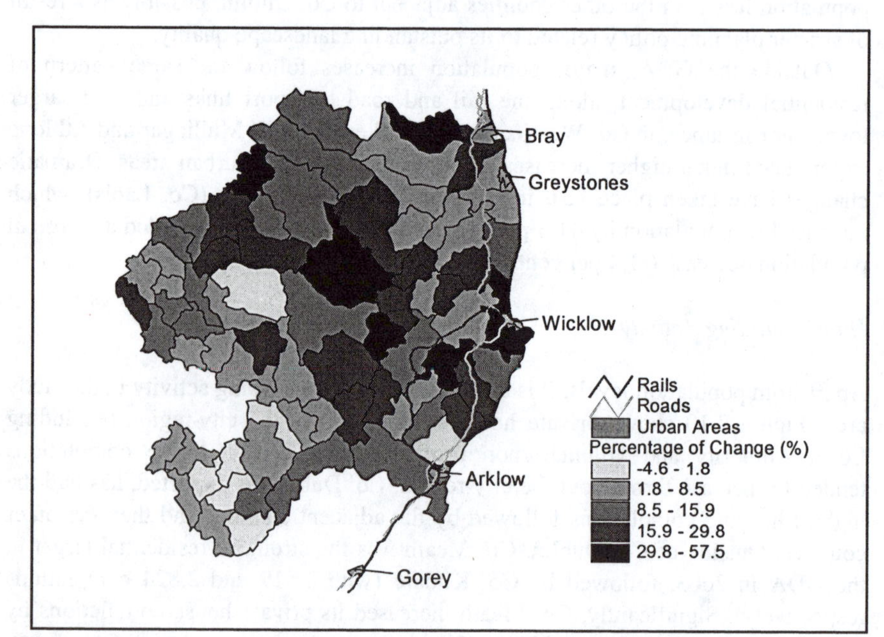

Figure 8.9 County Wicklow population changes, 1996-2002

In terms of housing type, a much higher proportion of new-detached housing development can be found outside of Co. Dublin, reflecting higher densities and increased apartment-living within the urban core. However, the availability of larger dweller units with individual front and back gardens may indeed be a major factor in migratory trends. It is unclear at present how much of this development outside Co. Dublin is comprised of dispersed single dwellings in the countryside, which requires further research. It appears that new detached house completions as a percentage of total completions peaked in most counties in 2000 or the year before. Significant differences from other types of houses are obvious for counties like Laois, Longford and Kilkenny. After 2000, the percentage of detached new houses started decreasing considerably in most counties and this trend continued until 2003.

Despite the percentage decreases in the numbers of new detached house completions, the actual numbers of detached houses as a whole, according to the latest census results, are very high. In most counties in the city-region, detached houses account for more than 50 per cent of all private households and accommodate more than 50 per cent of the people leaving in each county. In addition, most people in the study area own the house they are living in (Figure 8.11). In fact Ireland has one of the highest levels of home ownership in Europe

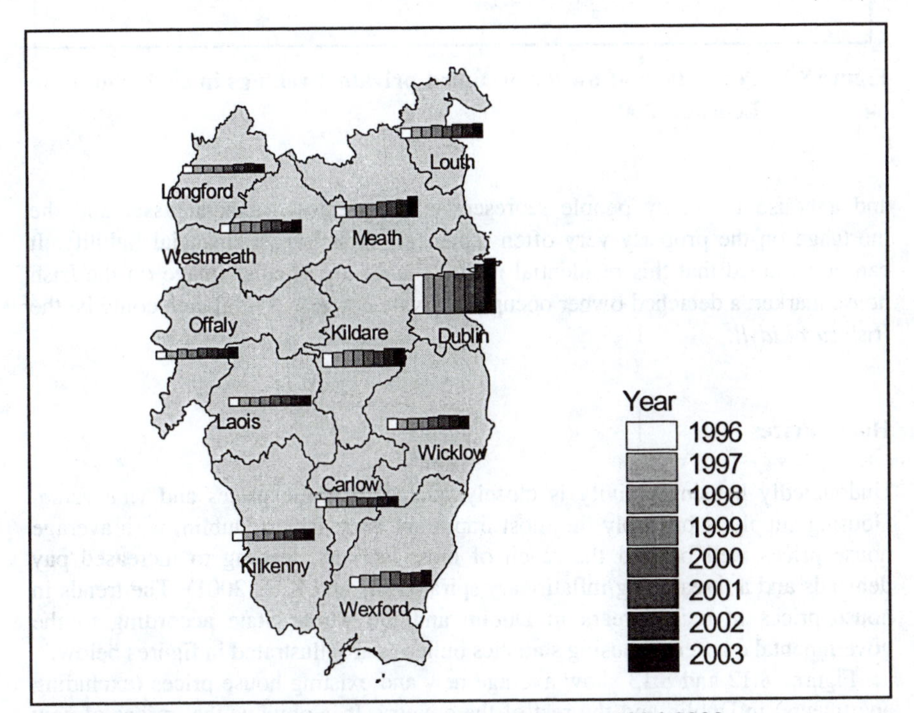

Figure 8.10 Private house completions in Leinster, 1996-2003

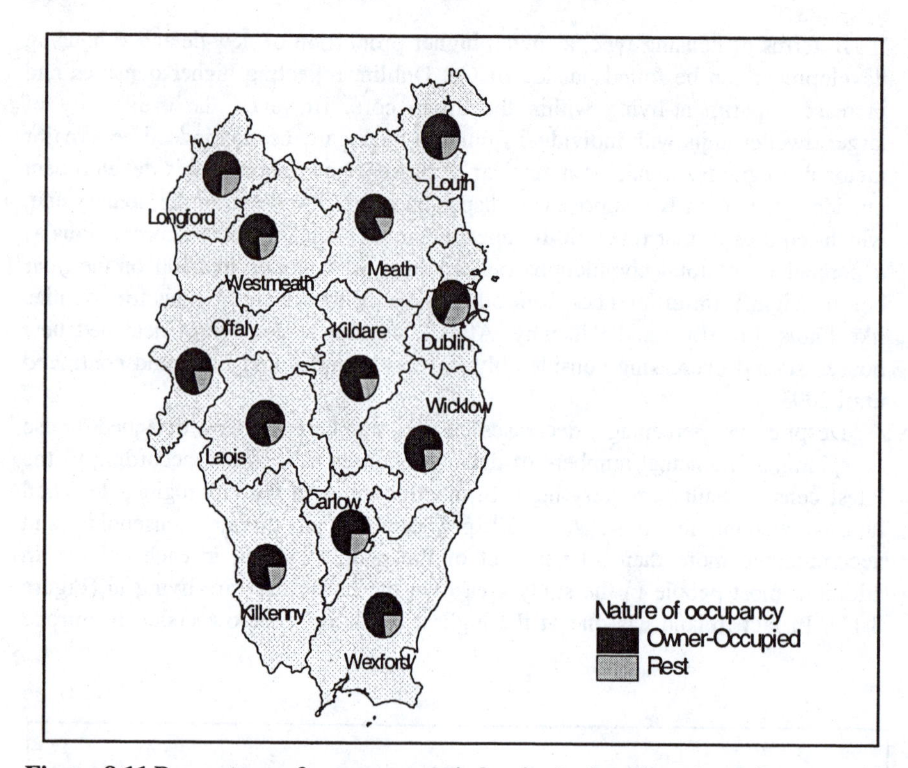

Figure 8.11 Percentage of owner-occupied private dwellings in each county in Leinster, 2002

and a house for many people represents their greatest financial asset and the mortgage on the property very often represents their largest financial liability. It can be assumed that this residential reality shapes a particular image on the Irish house market: a detached owner occupied private house in a rural area could be the Irish *rural idyll*.

House Prices

Undoubtedly housing activity is closely related to house prices and *vice versa*. Housing supply is probably the most important issue facing Dublin, with average house prices well beyond the reach of most workers, leading to increased pay demands and a threatening inflationary spiral (Ellis and Kim, 2001). The trends in house prices in recent years in Dublin and the whole State according to the governmental quarterly housing statistics bulletins are illustrated in figures below.

Figures 8.12 and 8.13 show average new and existing house prices (excluding apartments) in Dublin and the rest of the country. It is obvious that prices of new

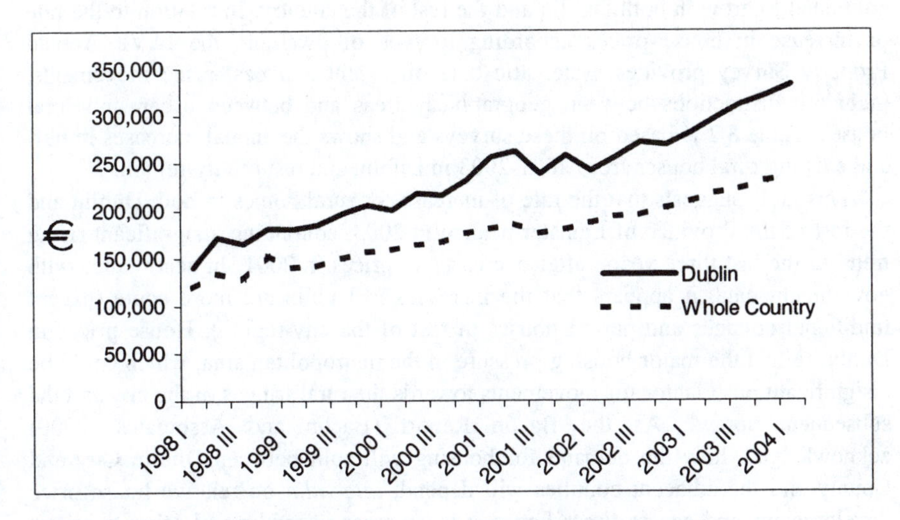

Figure 8.12 Average new house prices (excluding apartments)

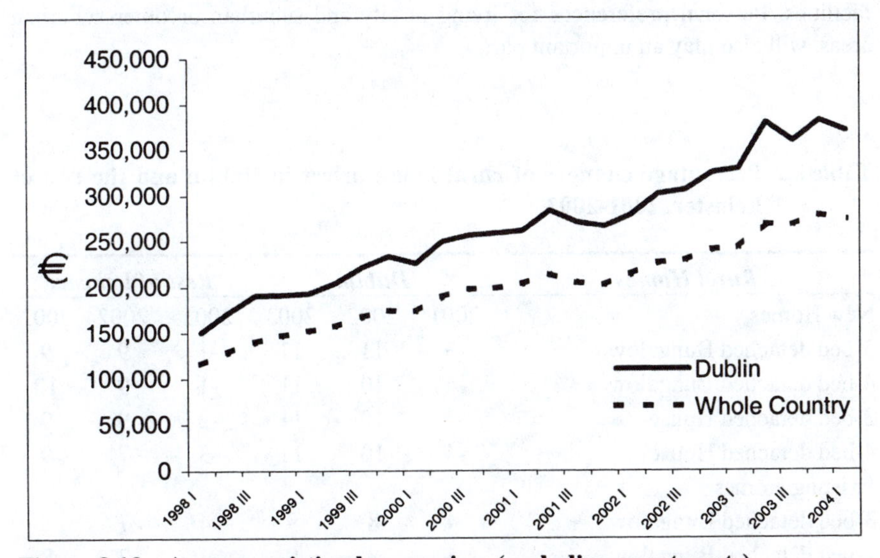

Figure 8.13 Average existing house prices (excluding apartments)

houses follow a constant increase and they are much higher in Dublin than the country as a whole. Despite the slowdown in the rate of increase after 1998, the level of prices illustrate a rising trend and as the third Bacon report[5] (Bacon and Associates, 2000) concludes this is an indication of continuing instability. Second hand houses appear to be more expensive than new houses and their prices have continued to grow in both Dublin and the rest of the country. In relation to the rate of increase in house prices according to type of dwelling, the IAVI[6] Annual Property Survey provides systematic data on recent and past price movements, including distinctions between geographical areas and between urban and rural houses. Table 8.2 is based on these surveys and shows the annual increases in new and existing rural houses from 2001-2003 in Dublin and rest of city-region.

This table suggests that the rate of increase for rural homes in both Dublin and the rest of the Province of Leinster peaked in 2003, continuing a significant rising trend in the last three years, after decreases in prices in 2001. In accordance with government data, it appears that the increases in Dublin are more acute (except traditional cottages and period houses in rest of the city-region). House prices in Dublin reflect the major housing pressure in the metropolitan area, which could be a significant push factor for movements towards the rural fringes of the city and the subsequent sprawl. As the Bacon Report (Bacon and Associates, 2000) acknowledged, how the demand for housing will split between Dublin City and County and the adjacent counties will depend, *inter alia* on relative house price developments and access times between these areas in public and other transport terms and relative endowments of social and recreational infrastructures and facilities. Personal preferences for living in city and suburban or outer suburban areas, will also play an important part.

Table 8.2 Percentage changes of rural house prices in Dublin and the rest of Leinster, 2001-2003

Rural Homes	***Dublin***			***Rest of Leinster***		
New Homes	2001	2002	2003	2001	2002	2003
3-bed detached Bungalow	-3	11	11	-1	9	9
4-bed detached Bungalow	-4	10	11	-1	9	10
3-bed detached House	-3	10	11	-2	9	9
4-bed detached House	-4	10	11	-3	7	9
Existing Homes						
3-bed detached Bungalow	-4	8	9	-4	7	8
4-bed detached Bungalow	-5	8	9	-3	7	8
3-bed detached House	-1	6	9	-4	7	8
4-bed detached House	-3	6	10	-3	6	8
Traditional cottages	-	5	10	0	11	10
Period houses on good grounds	-	9	10	2	11	11

Source: IAVI Annual Property Surveys (2001-03)

Travel-to-Work Commuting Patterns

A further issue of urban-rural migration is the relationship between commuting patterns and urban sprawl. For people living in the rural fringe of towns and work in urban areas, commuting is probably the most important trip of the day. Car ownership in the Republic of Ireland has grown rapidly in line with the economic performance, placing Ireland amongst the most car-dependent societies in Europe. There were almost 1.2 million private cars in the country in 1998 and car ownership exceeded 50 cars per 100 persons for the first time in the same year (Goodbody Economic Consultants, 2000). The rising level of car ownership has led to an increasing reliance on the private motor vehicle as the preferred mode of transport (Clinch et al., 2002). Indeed, according to the 2002 census, the private car appears to be the most popular mode of transport in the city-region. More than 50 per cent of people are commuting to work, school or college either by driving or as passengers in a car. The use of public modes of transport (bus and train) accounts for less than 40 per cent of the people living in the city-region whereas cycling and walking seem to be the least popular ways. More than 50 per cent of people are commuting to work, school or college by car (Figure 8.14).

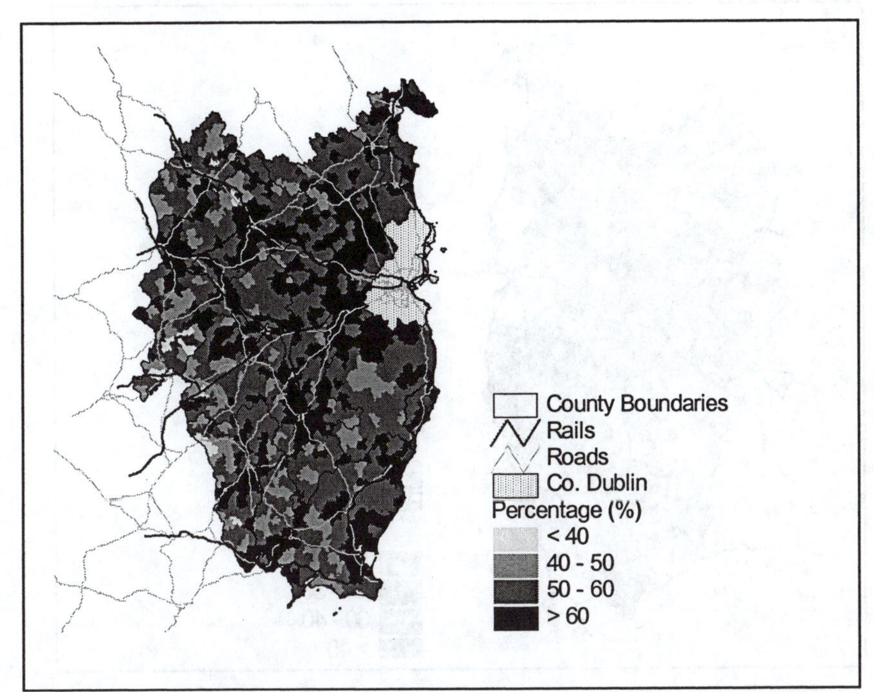

Figure 8.14 Percentage of people who travel to work, school or college by car (driving or passengers)

Figure 8.15 reveals the percentage of people who travel more than 15 miles (around 24 km) from home to work, school or college. It appears that long-distance commuting has became particularly characteristic in the GDA. Longer travelling distances also appear around urban areas like Waterford and Wexford. These are areas that significantly increased their population during the 1996-2002 period. According to Clinch et al. (2002), long distance commuting has led to further congestion, increased travel time to work, rising frustrations and stress, and increased fuel and associated pollution.

Conclusion

In summary, recent migration flows in the Dublin city-region generally occurred outside urban areas and towards the rural fringe. Urban-rural migration appears to follow the pattern of road and rail transport, which suggests a rural living preference but with the benefits of proximity to urban areas. In particular, the mid-eastern counties of the Greater Dublin Area have become significant residential targets with increasing private house completions (Co. Meath

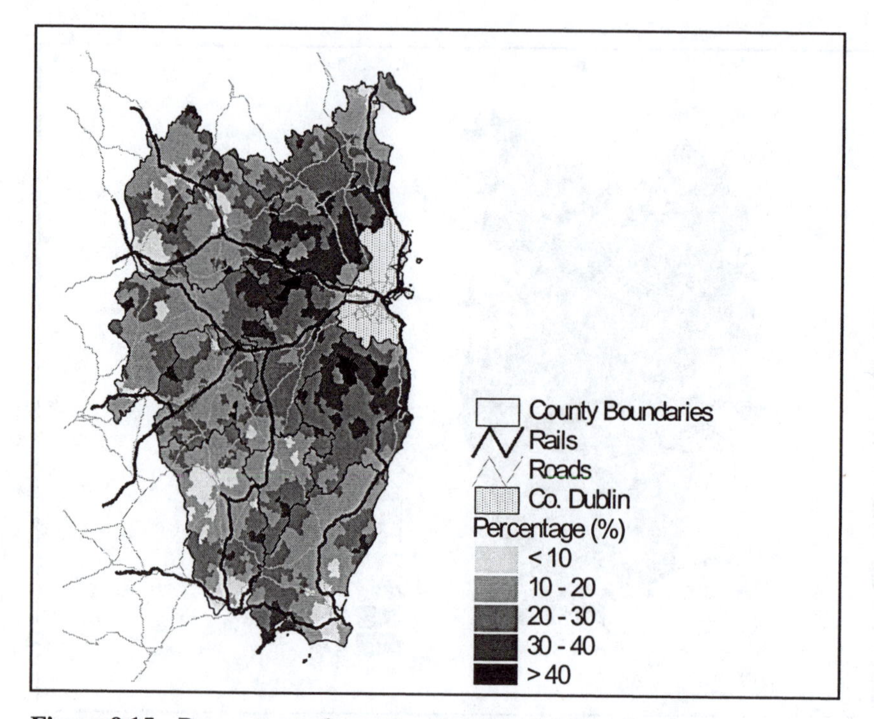

Figure 8.15 Percentage of people who travel 15+ miles from home to work, school or college in Leinster

increased its housing completions by 163 per cent in only 5 years whereas at the same time Co. Dublin increased its housing completions by 37 per cent). These trends are consistent with notions of counterurbanisation and metropolitan dispersal. In particular, the housing pressure in Dublin suggests a displaced-urbanisation (following Mitchell's classification) whereby people migrate to the countryside simply because it is cheaper to buy or build a house there. On the other hand, preferences for rural living in Ireland suggest the existence of an Irish rural idyll. An isolated house in the open countryside with the benefits of proximity to the capital or other urban areas appears to be a strong choice for city dwellers, facilitated by a relaxed planning regime. These urban dwellers' intentions or actions fit closely with ex-urbanisation and/or anti-urbanisation movements.

Lindgren (2002) observed that although a major focus of research has been to identify periods of population concentration or dispersal, primarily in advanced capitalist societies, scant attention has been paid to studying the individuals and households who decide whether or not to make a counterurban move. The NSS vaguely acknowledged that "some persons from urban areas seek a rural lifestyle with the option of working in and travelling to and from, nearby larger cities and towns" (p. 106). However, much less clear is the types of people this includes; the driving forces (push and pull) behind such relocations; and the extent that urban households in Dublin have preferences for a rural residential environment. Since urban-rural migration is not only a socio-economic but also a geographically selective process (van Dam et al., 2002), further baseline information is essential to provide an evidence-based approach to policy-making – for example, where do these migrants go and how much are they willing to trade the amenities of a rural residential environment with longer daily journeys to work and to nearby urban centres for services? What are the impacts of urban-generated rural housing in relation to local services, population levels and community cohesion? Through understanding consumer and lifestyle factors, in particular how these may vary through an individual's life cycle, consumer decision-making could be influenced more effectively by policy-makers to achieve desired policy outcomes. Moreover, although policy prescription currently advocates a compact city approach for the urban core and concentrated development in key settlements in the hinterland, evidence suggests that deconcentrated and decentralised spatial patterns are emerging. In this context, further research is also required to evaluate the effectiveness of existing and potential alternative policies and policy instruments.

Notes

[1] For the purposes of this chapter, the Dublin city-region is defined as the Province of Leinster, comprised of the following counties: Carlow, Dublin, Kildare, Kilkenny, Laois, Longford, Louth, Meath, Offaly, Westmeath, Wexford,

Wicklow. This definition extends beyond the Greater Dublin Authority administrative boundary, reflecting the increasing influence of the capital city.
[2] The term *counterurban* was proposed to describe a *pattern* of population distribution that is deconcentrated (small number of people being distributed in many settlements), and the term *counterurbanising* to describe the *process* whereby a settlement system is transformed from a concentrated to deconcentrated state (Mitchell, 2004).
[3] The survey was based on interviews with a probability based random sample of Irish adults (aged 18+) population (UII, 2001).
[4]The term Electoral Division was changed on 24 June 1996 (Section 23 of the Local Government Act, 1994) from District Electoral Division. There are 3,440 Electoral Divisions in the State.
[5] The Bacon Reports (three so far) were commissioned by the Government (Department of Environment and Local Government) for in-depth research on the national housing market.
[6] The Irish Auctioneers & Valuers Institute (IAVI) represents over 1,550 real estate agents and auctioneers in Ireland.

References

Aalen, F. (1997) 'The challenge of change', in Aalen, F, Whelan, K. and Stout, M. (eds) *Atlas of the Irish rural landscape*, Cork: Cork University Press.

Bacon, P. and Associates (2000) *The housing market in Ireland: An economic evaluation of trends and prospects*, Dublin: DOEHLG.

Berry, B.J.L. (1976) 'The counterurbanization process: Urban America since 1970', in Berry, B.J.L. (ed.) *Urbanization and counterurbanization*, Beverly Hills, London: Sage Publication.

Boyle, P., Halfacree, K. and Robinson, V. (1998) *Exploring contemporary migration*, UK: Longman.

Champion, A.G. (1989) 'Counterurbanization: the conceptual and methodological challenge', in Champion, A.G. (ed.) *Counterurbanization: The changing pace and nature of population deconcentration*, London: Edward Arnold.

Champion, A.G. (1998) 'Studying counterurbanisation and the rural population turnaround', in Boyle, P., and Halfacree, K. (eds) *Migration into rural areas, theories and issues*, Chichester: John Wiley & Sons.

Clinch, P., Convery, F. and Walsh, B. (2002) *After the Celtic Tiger, challenges ahead*, Dublin: O'Brien Press.

CSO (Central Statistics Office) (1979-2002) *Various Censuses*, Dublin: Stationary Office.

van Dam, F., Heins, S. and Elbersen, B.S. (2002) 'Lay discourses of the rural and stated and revealed preferences for rural living. Some evidence of the existence of a rural idyll in the Netherlands', *Journal of Rural Studies*, Vol. 18, pp. 461-476.

DOEHLG (Department of the Environment, Heritage and Local Government) (1997-2004) *Housing Statistics Bulletin*, Dublin: Stationary Office.

DOEHLG (Department of the Environment, Heritage and Local Government) (2004) *Sustainable Rural Housing, Consultation Draft of Guidelines for Planning Authorities*, Dublin: Stationary Office.

DOELG (Department of Environment and Local Government) (2002) *The National Spatial Strategy 2002-2020, People, Places and Potential*, Dublin: Stationary Office.

Dublin Regional Authority and Mid-East Regional Authority (2004) *Regional Planning Guidelines Greater Dublin Area 2004-2016*, Dublin: Regional Planning Guidelines project office.

Ellis, G. and Kim, J. (2001) 'City profile: Dublin', *Cities*, Vol. 18(5), pp. 355-364.

Gallent, N., Shucksmith, M. and Tewdwr-Jones, M. (2003) *Housing in the European countryside, rural pressure and policy in Western Europe*, London: Routledge.

Goodbody Economic Consultants (2000) *Transport demand*, Dublin: DOELG.

Gorton, M., White, J. and Chaston, I. (1998) 'Counterurbanisation, fragmentation and the paradox of the rural idyll', in Boyle, P., and Halfacree, K. (eds) *Migration into rural areas, theories and issues*, Chichester: John Wiley & Sons.

Gray, J. (2000) 'The Common Agricultural Policy and the re-invention of the rural in the European Community', *Sociologia Ruralis*, Vol. 40, pp. 30-52.

Hadjimichalis, C. (2003) 'Imagining rurality in the new Europe and dilemmas for spatial policy', *European Planning Studies*, Vol. 11(2), pp. 103-113.

Halfacree, K. (1997) 'Contrasting roles for the post-productivist countryside, a postmodern perspective on counterurbanisation', in Cloke, P. and Little, J. (eds) *Contested countryside cultures, otherness, marginalization and rurality*, London: Routledge.

Halfacree, K. and Boyle, P. (1998) 'Migration, rurality and the post-productive countryside', in Boyle, P., and Halfacree, K. (eds) *Migration into rural areas, theories and issues*, Chichester: John Wiley & Sons.

Halliday, J. and Coombes, M. (1995) 'In search of counterurbanisation: Some evidence from Devon on the relationship between patterns of migration and motivation', *Journal of Rural Studies*, Vol. 11(4), pp. 433-446.

IAVI (Irish Auctioneers & Valuers Institute) (2001-2003) *Annual Property Survey*, accessed at www.iavi.ie (5/10/2003).

Kontuly, T. (1998) 'Contrasting the counterurbanisation experience in European nations', in Boyle, P., and Halfacree, K. (eds) *Migration into rural areas, theories and issues*, Chichester: John Wiley & Sons.

Lindgren, U. (2002) *Counterurban migration in the Swedish urban system*, Centre for Regional Science (CERUM) Working Papers 57, Sweden: Umeå University.

McGrath, B. (1998) 'Environmental sustainability and rural settlement growth in Ireland', *Town Planning Review*, Vol. 3, pp. 227-290.

Marsden, T. (1999) 'Rural futures: The consumption countryside and its regulation', *Sociologia Ruralis*, Vol. 39, pp. 501-520.

Mitchell, C.J.A. (2004) 'Making sense of counterurbanization', *Journal of Rural Studies*, Vol. 20, pp. 15-34.

Nivalainen, S. (2003) 'Who moves to rural areas? Micro-evidence from Finland', Paper presented to the *European Regional Science Association (ERSA) Conference*, Finland: Jyväskylä.

Perry, R., Dean, K. and Brown, B. (1986) *Counterurbanisation: International case studies of socio-economic change in rural areas*, Norwich: Geo Books.

Spatial Planning Unit (2001) *Rural and urban roles – Irish spatial perspectives*, Dublin: DOELG.

Stead, D. (2002) 'Urban-rural relationships in the west of England', *Built Environment*, Vol. 28(4), pp. 299-310.

UII (Urban Institute Ireland) (2001) *Public attitudes to the environment and other issues*, Research and Evaluation Services, Dublin: UCD.

PART II
SUSTAINABLE COMMUNITIES

Chapter 9

'We're Too Busy for That Kind of Stuff': Progress Towards Local Sustainable Development in Ireland

Geraint Ellis, Brian Motherway and William J.V. Neill

Introduction

In the years surrounding the Rio Summit of 1992, the apparition of sustainable development appeared on the international stage and was warmly greeted by the majority of the world's governments. It has remained a prominent objective ever since. The ambiguity of the term is infamous and has been the motivation for both its widespread adoption and the frustration over to its progress. There is plenty, perhaps too much, debate around what sustainability or sustainable development implies, which has been usefully disaggregated by Jacobs (1995) in his suggestion that we should conceive of sustainable development on two levels. The first is the level of vague principle, akin to freedom and democracy, that is almost universally agreed (although not by all, see for example, Beckerman, 1994) and the second a level of interpretation or implementation, where there is a high degree of contestation. The first of these levels is potentially less problematic, providing the 'vision-thing', which has, on occasion, acted as a powerful energising force for the adoption of national and regional sustainability strategies. The second level is where the sustainability paradigm faces it's key challenges as it is here that difficult trade-offs have to be made and where opposition from vested economic and ideological interests have to be confronted. Despite these problems, it is clearly only through implementation that sustainable development can begin to prove its worth and display its long-term relevance.

Almost every level of governance across the island of Ireland has signed up to a vision of sustainability, suggesting it is a paradigm of some importance. However, when we scratch the surface, there appears to be significant difficulties in reforming and reframing decision-making in a way that encourages the transition towards sustainable development. These problems appear most acutely at the local level, where constraints on implementation have become most sharply focussed and where the promise of a sustainable future can be best tested. The viability of local models of sustainable development is fundamental, as it is here that key consumption-informing services such as land use planning and environmental health are delivered and where the potential of sustainability to transform patterns of governance is best realised. Shifts in governance are central to the transition to sustainable development (e.g. O'Riordan, 2002; Selman and Parker, 1999 and Barry, 1996) and should include a redefinition of the ethics of decision-making and a widening of our 'communities of concern'. This implies, among other things, a deepening of inter- and intra-generational empathy, an acceptance of cross-border responsibilities and an increased accumulation of social capital – all issues that have come under increased pressure in the Republic of Ireland and Northern Ireland as a result of rapid economic growth, increased social tension and in the North, the continuing entrenchment of political and religious division. It therefore follows, that if Ireland, North and South, were to adopt the goal of local sustainable development, it would not only allow the island to contribute more positively to global issues such as climate change and resource depletion, but could offer a means of enhancing community development (Warburton, 2002). This would appear to be a desirable objective, which could contribute to a range of other goals from crime reduction to educational attainment, while also enhancing cooperation across the Irish border. Cross-border cooperation on sustainability can thus be justified, not only for managing the single ecological unity of the island, but also in terms of the aspirations for greater interaction and mutual understanding. As such, sustainable development has been identified by the Environment Sector of the North/South Ministerial Council as a topic of research, specifically the need for the 'Identification of strategies and activities which would contribute to a coherent all-island approach to the achievement of sustainable development' (North-South Ministerial Council 2000).

This inevitably raises questions over the extent to which local sustainable development has taken hold on the island of Ireland, its impact on community values and whether there is potential in pursuing local sustainability as an all-Ireland strategy. This chapter attempts to probe some of these issues, first by setting out the broader context for local sustainability, followed by a description of the relevant policy context either side of the border and then goes on to evaluate the degree of implementation in Northern Ireland and the Republic of Ireland. Finally, it considers the implications for cross-border sustainability strategies and how local sustainability could be related to the issue of citizenship. Much of this chapter is drawn from a study of local sustainable development undertaken for the Centre for Cross-Border Studies (Ellis et al., 2004).

Local Agenda 21: Sustainability and the New Localism

A dominant theme in debates over sustainability is an increased emphasis on action at a local level, summed up in the maxim 'think globally and act locally'. Indeed it has been suggested that 60 per cent of agreements made at the 1992 Rio Summit and 40 per cent of the European Environmental Action Plans have to be implemented at the local level (Gilbert et al, 1996). This has contributed to a 'New Localism' (Marvin and Guy, 1997) where 'local' is conceived in both physical (i.e. spatial, ecological and administrative systems) and social terms (i.e. reflecting common bonds of culture, identity and tradition) (Selman, 1996). Such a view argues that local authorities are pivotal to sustainability because they control key public services (e.g. waste management, transport, housing and education); are major local consumers of resources themselves; and are therefore in a good position to adopt management systems and energy conservation (Gilbert et al., 1996). As the lowest tier of government, local authorities are also best placed to reflect local political priorities, provide democratic accountability and engage citizens and local stakeholders in the dialogue needed to precipitate sustainable development (Glass, 2002; Selman and Parker, 1997). Although the local arena is neither simple nor exclusive, the potential it offers for effecting long term change has been recognised in global agreements for sustainable development, most notably that of Local Agenda 21 (LA21).

From Local Agenda 21 to Local Action 21

The origins and scope of LA21 are generally well known (see Brugman, 1997; Hom, 2002), becoming established in Chapter 28 of Agenda 21, the agreement signed by over 150 countries at the Rio Summit in 1992 (United Nations, 1993). LA21 essentially calls for local authorities to enter into a dialogue with their citizens, in order to encourage local action and the cultural and behavioural change required for sustainable development. This offered an opportunity to interpret and relativise the Rio action plan according to local conditions, taking into account existing programmes, social and economic and demographic trends. O'Riordan (2002) suggests that if the wording of the initial Rio agreement is taken literally, LA21 could be interpreted as narrowly as just promoting local authorities' own eco-auditing, recycling and energy efficiency efforts. However, LA21 has come to represent far more than this, so that according to Lafferty (2001) it has become expected that an LA21 process would entail:

- An explicit attempt to relate environmental effects to economic and political pressures.
- An active effort to relate local issues and decisions to global impacts.
- A focused policy for achieving cross-sectoral integration of concerns, goal and values.
- Greater efforts to increase community involvement, including multi-sectoral engagement, participatory assessment and target setting.
- A commitment to define local problems within broader ecological, geographical and temporal frameworks.
- Specific identification with the Rio Summit and Agenda 21.

The potential of LA21 has been widely recognised, with 6,500 local authorities in

113 countries having established such a process by 2001 (ICLEI, 2002). The initiative has been subject to extensive monitoring (eg Lafferty and Eckerberg, 1998a; Lafferty, 2001; ICLEI, 2002) with numerous accounts of the LA21 experience in specific communities (eg Selman and Parker, 1999; Kelly, 2000; Moser, 2001; Sharp, 2002 and Ryan, 2002). These accounts further underline the significance of the initiative, highlighting the benefits of LA21 as a way of inculcating local sustainability.

Despite the potential and initial enthusiasm for LA21, by the end of the 1990s interest in the initiative began to wane, particularly in those countries such as the UK and Sweden that had experienced early widespread adoption of LA21 (Lafferty and Coenen, 2001). After ten years of momentum, there thus appeared to be a growing sense of 'LA21 fatigue' (Otto-Zimmerman, 2002) as councils and stakeholders became increasingly frustrated with progress, particularly in relation to implementation issues. This led to attempts to reinvigorate the drive towards local sustainability at Rio+10, the World Summit on Sustainable Development (WSSD), held in Johannesburg in 2002. This recognised that LA21 had been an effective tool for raising awareness of sustainable development and the role of communities in its delivery, but a decade on, priority should shift more firmly to securing sustainability in practice. This was agreed in the form of Local Action 21 (LAn21), a framework for the post-Johannesburg decade of LA21 that was proposed as a motto, mandate and movement for accelerated implementation of sustainable development (Otto-Zimmerman, 2002). Although it is still a little early to fully evaluate the impact of this renewed commitment, it is notable that there has only been limited post-Summit discussion on LAn21 and with practically no national debate within the UK and the Irish Republic, there remains major questions over the prospects for local sustainable development on the island of Ireland.

Sustainability Policy on the Island of Ireland

Against the backdrop of the international aspirations for LA21, the national governments of the Republic of Ireland and in the case of Northern Ireland, the UK government, have made some attempt to frame strategic approaches to sustainable development and specifically to encourage local action on this issue. A detailed account of the broader policy contexts in the Republic of Ireland and Northern Ireland are provided elsewhere (Ellis et al., 2004), but are briefly summarised below.

Northern Ireland

As part of the UK, Northern Ireland has been expected to meet the aspirations of successive national sustainable development strategies, beginning with This Common Inheritance (DoE [UK], 1990), to the more current vision, A Better

Quality of Life (DETR, 1999a) and the accompanying national sustainability indicators (DETR 1999b, DEFRA 2003). The UK government has also been relatively successful in integrating a whole range of other policy concerns with the sustainability agenda, including planning, transport, social inclusion and international cooperation. While this provides a fairly robust national framework for implementing and measuring the progress towards sustainable development, unlike other devolved administrations of the UK, Northern Ireland has done little to develop its own regional sustainability priorities. Although the Department of the Environment published a rather tentative discussion paper in May 2002 (DoENI, 2002), at the time of writing (2004), there been no further development and there is little evidence that this will ever emerge as a coherent strategy that could frame the regional transition to sustainable development. This does not mean, however, that sustainability is completely neglected in the policy agenda in Northern Ireland, as it was included as a theme in the Programme for Government (NI Executive, 2001) of the NI Assembly and reflected in a number of other strategic policy documents such as the Regional Development Strategy (DRD, 2001), Northern Ireland Biodiversity Strategy (EHS, 2002), and Investing in Health (DHSSPS, 2002) among others. These all provide a loose commitment to sustainable development, but do so in the absence of any coordinating framework.

Republic of Ireland

National sustainable development policy in the Republic of Ireland has witnessed a similar evolution to that of the UK, albeit in a more compressed timescale, beginning with the State of the Environment Report (EPA, 1996), shortly followed by the National Sustainable Development Strategy (DoE [RoI], 1997). This strategy has provided the guiding framework for all policy that has followed, including that relating to LA21 and to other policy areas, such as poverty, (DSFA, 2002), waste management (DoELG, 2002b), climate change (DoELG, 2000), biodiversity (DAHGI, 2002) and spatial development (DoELG, 2002c). The Department of the Environment and Local Government (DoELG) undertook a comprehensive review of the implementation of sustainable development in Ireland prior to the WSSD, which highlighted the links between the various policy areas and overall progress on integration (DoELG, 2002a). The review sets out the priorities for the next ten years and is comprehensive in the coverage of the types of sustainability initiatives instigated at the national level.

This national policy framework plays a critical role in enabling and encouraging local activity in sustainable development either side of the Irish border. In an international context, progress on LA21 has been reported to be relatively successful in the UK as a whole, where it is seen as one of the national pioneers in Europe, with an early take up by many local authorities (Lafferty and Eckerberg, 1998; Lafferty, 2001). The initial progress of LA21 across the UK has been attributed to the example set by a number of trailblazing local authorities such as Sutton, Leicester and Lancashire and to the support of the Local Government Management Board (Church and Young, 2000). It was given a further boost in 1997, when the Prime Minister expressed a need for every UK local authority to have completed an LA21 plan by December 2000 (LGMB, 1999), supported by further central government guidance on the development of LA21 strategies (DETR, 1998). As a result, the UK has seen relatively high levels of public involvement in local sustainability debates, reflected in the fact that at least 91 per cent of local authorities have agreed an LA21 action plan (Tuxworth, 2001). It does not appear, however that this national experience is being effectively replicated in Northern Ireland where, in the virtual absence of regional government support for LA21, promotion of it has been largely left to various voluntary sector bodies, such as Bryson House, Sustainable NI and Northern Ireland Environmental Link. This situation has been further frustrated by the institutional arrangements for local government in the province that means its twenty-six District Councils have only a narrow range of responsibilities (Knox, 1998). Although these governance arrangements are currently under review (NI Executive 2003) and have been supplemented by the formation of Local Strategic Partnerships (LSPs), it means that compared to most European municipalities, Northern Irish local authorities have reduced financial resources, limited development capacity and less ability to integrate key services, such as housing, transport and planning into sustainability initiatives.

Within the Republic of Ireland, national progress on implementing LA21 has been perceived to be less advanced than the UK. Indeed, the same assessment that described the UK as a pioneer of LA21, suggested the Irish Republic's national performance was one of the most 'tardy' in Europe (Lafferty and Eckerberg, 1998; Lafferty, 2001). Mullally (2001) has described progress as a series of policy milestones that includes the publication of initial Guidelines on Local Agenda 21 (DoE[RoI], 1995), the National Sustainable Development Strategy (DoE [RoI], 1997), revised LA21 guidance (DoELG, 2001) and the review of progress towards sustainable development, prepared in advance of the WSSD (DoELG, 2002a). The evolving policy context has been accompanied by a programme of institutional modernisation, primarily under the banner of Better Local Government (DoE [RoI], 1996), which has enhanced the capacity of councils to deliver on sustainability. Particularly significant has been the establishment of inter-sector County/City Development Boards (CDBs), which offer enhanced potential for brokering local consensus on economic, social and environmental issues. As a result of these developments, Mullally (2001) suggests that LA21 in the Republic of Ireland largely remains at the level of 'latent potential' rather than making 'visible progress'. He does however, suggest that there are grounds for optimism in that an institutionalising process towards local sustainable development has slowly been taking place, thus enabling some of the structural impediments (eg. central government control of local authorities and poor participative processes) to be overcome (Mullally, 2001).

Progress and Status of Local Sustainable Development on the island of Ireland

While there is no shortage of reviews and evaluations of LA21 across the UK (eg Audit Commission, 1997; Voisey, 1998; Buckingham-Hatfield and Percy, 1999; Tuxworth 2001; Young, 2001), comparative information for Northern Ireland and the Republic of Ireland is relatively scarce. The last available data for Northern Ireland, from 2000, suggested that only nine out of twenty-six councils had adopted LA21 strategies. For the Republic of Ireland, Mullally (2001) suggested that only 20 local authorities had begun to engage in LA21, a disappointing picture confirmed by other assessments (eg Comhar, 2001; ESI, 2002).

The authors updated this data as part of a major review of implementation issues related to local sustainable development in Northern Ireland and the Republic of Ireland (Ellis et al., 2004). This took the form of surveys with local authorities and social partners, a series of focus group discussions with those involved in local sustainable development and analysis of four case studies of projects that specifically explored the links between governance and sustainability. This multi-dimensional approach resulted in a rich data set that reveals a number of connections between local sustainable development and citizenship. Prior to discussing these issues, it is necessary to provide a more general picture of local sustainable development either side of the Irish border.

Progress of LA21

Although LA21 has been championed as the key initiative for the promotion of local sustainable development, ten years after its launch, only 54 per cent of local authorities on the island of Ireland (58 per cent in Northern Ireland and 50 per cent in the Republic of Ireland) have even begun a process they consider to be LA21, with 16 per cent reporting that they have no intention of ever developing such an initiative. This compares poorly to other areas in Europe and the UK, where in some examples (eg Scotland and Sweden) all councils are reported to have adopted LA21. There is, of course, a difference in a quantitative evaluation of those councils that have initiated a process and a qualitative assessment of what has actually been achieved. Here, again the island of Ireland appears to be performing poorly, with very few council areas (14 per cent) having actually started to implement such a policy (see Figure 9.1) and low levels of innovation, particularly in terms of the efforts made to engage communities and the roles played by local stakeholders. The content of the LA21 processes tend not to reflect the integrative ideal of sustainability, being dominated by environmental issues, with only a minority of social issues and an absence of economic issues (see Figure 9.2), as noted by one of the focus group participants: '...in local government the agenda for development is absolutely 'Concrete is Good' ... by raising social issues you literally feel ostracised' (Local Authority Officer, Dublin focus group).

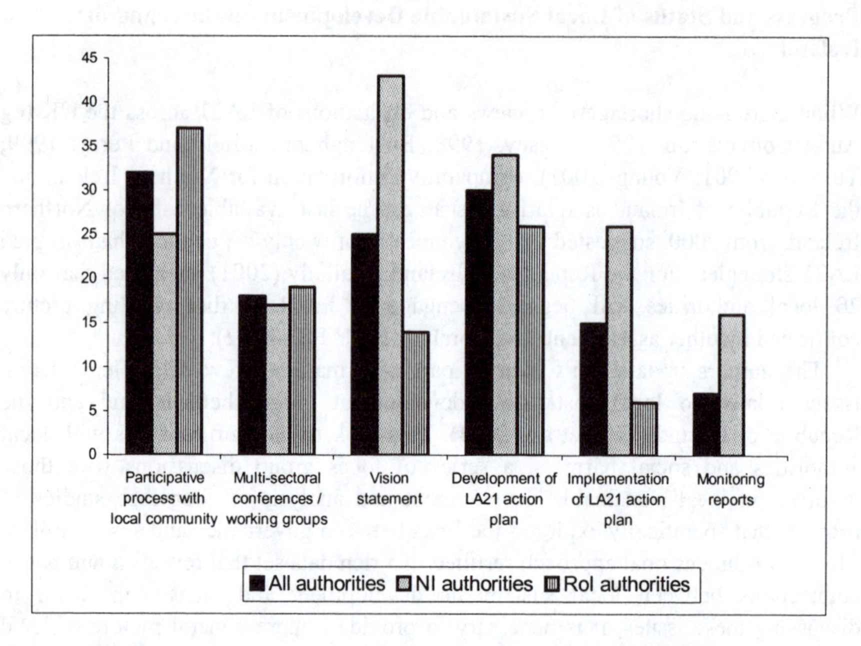

Figure 9.1 Percent of councils actively engaged in LA21 policies

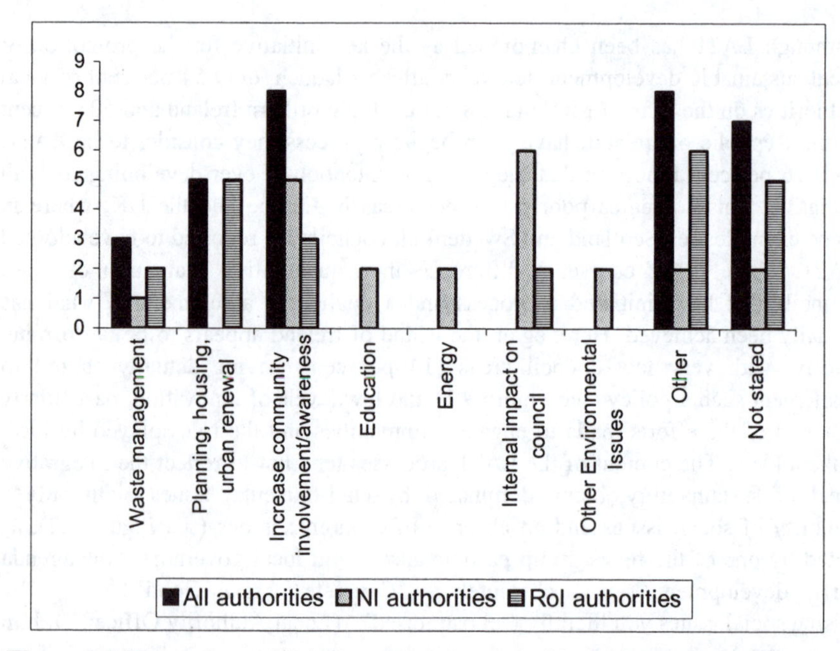

Figure 9.2 Dimensions of LA21 strategies engaged in by local authorities

The findings are not, however, entirely pessimistic as the research uncovered a wide range of examples of successful local sustainable development, which have not only focused on establishing frameworks for addressing environmental issues and embedding sustainability concepts in local government, but also provided some path-finding examples of participatory inclusion. It is also clear that LA21 has been successful in raising awareness of the possibilities of local sustainable development and stimulated debate on how local areas can contribute to the challenges created by Ireland's links to a global community facing severe ecological, economic and social problems. Indeed, some of the more optimistic participants in the focus groups, took a longer-term view, seeing the transition towards sustainability having at least begun and acknowledging the potential length of such a journey:

> Do we really think we are going to produce sustainability? … of course not, perfection is never going to happen, but we do need to accept the good that has happened and not be disappointed that we have not achieved 100 per cent, if we achieve 5 per cent, that's a huge achievement…. Although I'm not saying we've even achieved that yet (Participant, Local Authority Officer, Cork focus group).

However, overall the research suggests that progress towards and achievement of local sustainable development is disparate across the island of Ireland. This has resulted in a high level of frustration amongst social partners and local authority officers, leading some to question the adequacy of LA21 as a way of securing the transition to sustainability. This tends to reflect the 'LA21 fatigue' found at the broader geographic scale with questions around the ability to move from debate to action being the key concern across Ireland. The reasons for this appear to fall into two broad categories – the coherence of the LA21 initiative itself and the capacity of local authorities to respond to the challenge of sustainable development.

Coherence and Relevance of LA21

It is important to understand the context in which LA21 was originally launched – where the concept of sustainable development was little known and even less understood. One of the primary objectives of LA21 was therefore to promote awareness of sustainability at the local level. Although it has achieved much in this respect, as the novelty of the initiative has worn off and some of its component principles mainstreamed, the direct relevance of LA21 appears to have declined. In particular, the adoption of sustainability as a goal by a wider range of public and private sector organisations appears to have diluted the coherence of LA21, particularly in the minds of social partners and local communities. The research attempted to gauge if local authority officers and social partners felt that LA21 was an appropriate description of the process of developing sustainability at a local level, with only 43 per cent of local authorities and 29 per cent of social partners across the island of Ireland agreeing it was. It is also important to note that only a proportion of the activity regarded as being local sustainable development is formally accounted for under the LA21 banner, with 75 per cent of all councils stating that they were involved in other initiatives, apart from LA21, that contributed to sustainable development.

This suggests a significant lack of confidence in the terminology that claims to represent the global consensus on forging local sustainable development and which is supposed to have guided local authorities actions in this area for the last 10 years. Discussion at the focus groups shed some light on this, confirming the low level of confidence in the current 'brand' of local sustainable development. Participants from a deeper ecological perspective articulated the view that LA21, and indeed the concept of sustainability itself, could sometimes simply be labels superficially applied to convince the community of the supposedly public interest virtues of policies and plans – the rhetorical spin of sustainability acknowledged by many other researchers (eg Myers and MacNaghten, 1998). As one of the focus group participants noted:

> ...one point of view has been to justify progress in terms of [economic] development, but just put the label of sustainability on it and people will think it is good for them (Participant, Social Partner, Cork focus group).

While it is also clear that LA21 and local sustainable development have a degree of international saliency (albeit perhaps limited to policy-makers), no convincing alternative terms were thrown up by either the survey or focus groups, with 'quality of life' perhaps the best contender, though far from ideal. Therefore, while there is some dissatisfaction about the adequacy of the existing language of sustainable development, there seem to be no viable alternatives and to those involved in the process at least, LA21 still retains some relevance (see Figure 9.3). This figure also illustrates the fact that it is those less close to policy development (ie social partners) who are more confused about the relevance of the term.

This therefore poses a difficult dilemma – the terminology of sustainable development is clearly inadequate and now even becoming an obstacle to the implementation of the values it embodies. As a result of this, a large majority of local authority officers, who have become accustomed to these semantic debates, believe that LA21 remains relevant to the overall objectives of sustainable development, as expressed at a focus group:

> We use terms like LA21 and sustainable development interchangeably, but most people in the council know what we mean...it identifies the broader agenda (Participant, Local Authority Officer, Dublin focus group).

On the other hand, the public and even social partners are significantly confused about this issue. There is therefore a substantial communication problem that needs to be addressed.

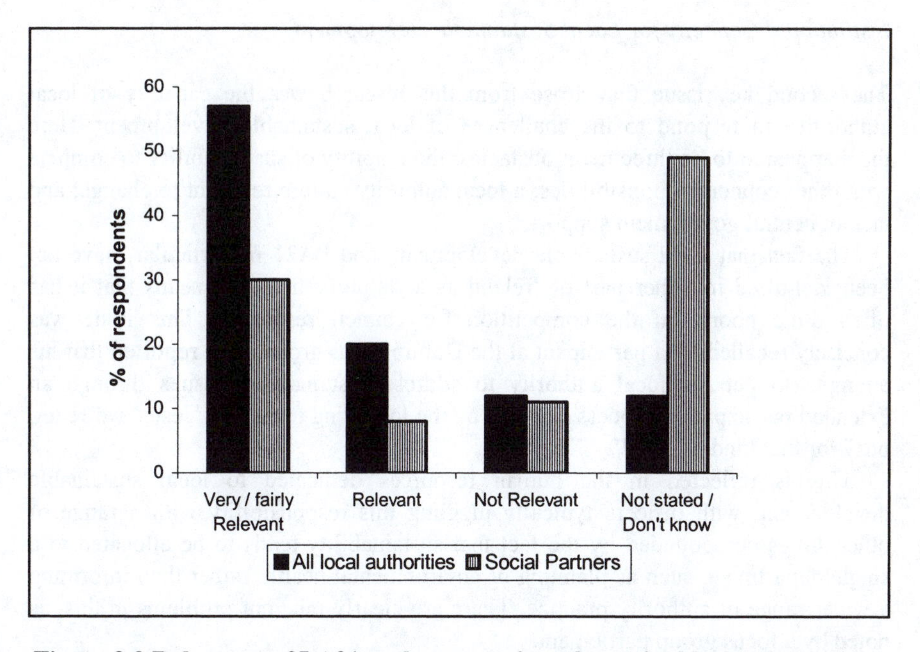

Figure 9.3 Relevance of LA21 to the promotion of sustainable development

It also appears evident that in the years since the Rio Summit, many activities that were then part of a campaigning agenda have now been incorporated into various and diffuse aspects of local authority operating cultures and institutional arrangements. As a result of such institutionalisation, the concept of sustainable development has lost some of its visionary appeal and the project has had to draw in an expanded set of stakeholders, with very different cultures and worldviews. It is therefore almost inevitable that the more idealistic elements have been replaced by more practical, management issues, which some have interpreted as reasons for LA21 having lost its energy and wider appeal. It is in a sense a victim of its own success. The following are examples of focus group comment in this regard:

> I am on the steering committee of LA21 in the city council and we have not met in a year. I really would say that LA21 as envisaged in 1992 is no longer there. We are doing it in another guise and a lot of other guises (Participant, Local Authority Officer, Dublin focus group).

> What you are finding is people do not always take actions under LA21 but there is actually a lot happening that would be classified as such. The problem is that there is not a broad understanding within communities and with people in the street as to what LA21 is. I think the name is unfortunate (Participant, Social Partner, Cork focus group).

Institutional Capacity for Local Sustainable Development

The second key issue that arose from this research was the capacity of local authorities to respond to the challenges of local sustainable development. Here there appeared to be three main obstacles; the inability of sustainability to compete with other council responsibilities; a local authority culture resistant to change; and lack of central government support.

The fact that local sustainable development, and LA21 in particular, have not been identified in either part of Ireland as a statutory function means that it has often done poorly in the competition for council resources. This issue was concisely recalled by a participant at the Dublin focus group, who reported that his attempts to get his local authority to address sustainability issues through an extended participation process was met by the following response: 'Jesus, we're too busy for that kind of stuff!'

This is reflected in the human resources dedicated to local sustainable development, with officers typically juggling this responsibility with a range of other duties compounded by the fact that sustainability tends to be allocated to a single department, such as planning or environmental health, rather than informing a wider range of authority practice. There are clearly inherent problems in this, as noted by a focus group participant:

> If you are given responsibility for LA21 or social inclusion it becomes a ghetto…there is a danger that people will say 'Oh, he's looking after that and the rest of us can get on with our lives… (Participant, Social Partner, Cork focus group).

This highlights the fact that the corporate culture of local authorities appears to be unreceptive to the principles of sustainability, so that it lacks prominence and integration with the range of council activities. Yet this cultural change could be fundamental to the ability to make significant progress on the issue, as noted in a focus group:

> It is a very slow process, because what you are trying to do is change a culture within an organisation, but that's the only way you will do it… you can have all the leaflets you like in the world… you can have all the projects you want, but unless people have this stuff in their subconscious, it's not going to happen (Participant, Local Authority Officer, Cork focus group).

It has also become clear, through survey and focus group evidence, that the single most important factor in successful examples of local sustainability initiatives was the support of senior council officers, as this was the key means of mobilising critical local authority resources. Indeed, the focus groups and case studies reveal a situation where institutional inertia is kept at bay largely through the energy and commitment of non-elected individuals, labelled as 'Practice-Orientated Evangelists' (POEs). The reliance on such individuals has major implications for the fragility and long-term viability of the local sustainable development agenda

North and South of the Irish border.

A final point here is the emphasis placed on the role of national government, which was identified in surveys as the most significant factor in limiting promotion of local sustainable development. The policy contexts for either side of the Irish border were briefly reviewed earlier, but this fact has been identified as being critical on a European level (ICLEI, 2002) and suggests that while it is relatively easy to point to failings by local authorities, the constraining, enabling and encouraging role (or lack of it) of central government must be acknowledged, as noted by one council officer:

> People must feel it makes a difference by fitting into national and international policies. We can do very valuable work locally with small projects but unless the national projects are analysed properly we are not going to have a major impact… (Participant, Local Authority Officer, Dublin focus group).

It is clear that there is a global imperative for moving towards more sustainable lifestyles and this has to have a key local dimension if it is to be effective. However the review of the progress of local sustainable development across Ireland finds a rather mixed picture, with islands of good practice amongst a sea of despondency and frustration. This appears to be the result of failures at several levels of governance and suggests that one of the prime requirements for further embedding of local sustainability concepts is re-energising and building the capacity of practitioners (including those in the voluntary and private sectors) through the support and demonstrable commitment of senior managers and government. Local sustainable development in Ireland also needs a renewed focus to lift it from LA21 fatigue – it is suggested here that one potential way of achieving this may be to develop a new 'island ethic of sustainability', primarily based on citizenship and cross-border collaboration.

Citizenship and Local Sustainability

By definition, sustainable development implies enhanced communal obligation and should enable people to better relate to each other's needs, with obvious implications for governance. This underlines that the path to sustainability is above all, a political process with the concept of citizenship at its core. As a consequence, it is possible to link local sustainability to the wider debates on the nature of citizenship (Isin and Turner, 2002) and the need to overcome cultural and political differences through the development of 'bridging social capital' (Putnam, 2000) – something that is being overtly applied in the US through 'civic environmentalism' (Agyeman and Angus, 2003). This is also implied by the fact that LA21 should be 'discursively created' rather than being an authoritatively given product (Barry, 1996, p. 116). Similarly Selman and Parker (1999) in their study of LA21 in England, found that the dominant 'storyline' that comes out of the process is not the traditional politics of inequality, but the 'emergent politics of identity, with its emphasis on participation and citizenship' (p. 56). From this perspective, sustainability could be linked to the reinvigoration of local democracy, which has been a policy goal of the government in the Republic of Ireland, while in Northern Ireland the extension of political reform from the regional level to the local is a logical next step in the ongoing Peace Process. Indeed, one focus group participant noted how the environment can be:

> ...common ground on which people can talk to one another and get an alternative conversation going...it can encourage people to look outwards rather than inwards (Participant, Social Partner, Northern Ireland focus group).

Indeed, while components of sustainable development are now being pursued discretely in a wide range of other areas of public policy, a critical point here is that on their own, these are not the same as more distinct approaches to local sustainable development, such as LA21, because in stressing one element of sustainability, such as democratic renewal, some of the diagnostic and important objectives of local sustainable development may be lost. Amongst these is the idea that local areas (and their citizens) can begin to see themselves as part of the global community, with subsequent reflections on issues of justice and ethical responsibility. Furthermore, while most recent governance reforms tend to enhance democracy and the effectiveness of public services, they tend to relegate the priority of limiting the adverse environmental outcomes of development, despite growing warnings of global consequences. It is argued, therefore, that local sustainability on the island of Ireland should be seen in its broader context, not just as being related to discrete areas of environmental protection or social inclusion, but as a way of extending democratic involvement that can foster common ethical concerns, which may in turn create a binding force in civic life. A key message should therefore be that the pursuit of local sustainable development is not only a virtue in itself, but also contributes to addressing some other major social concerns in western societies.

Conclusion: Towards an Island Ethic of Sustainability

The necessity of placing a renewed emphasis on the citizenship perspective on local sustainable development can be seen as being both acute and apposite in an Irish context. First is the fact that both parts of the island currently lack sufficient 'civic glue' to tackle their key social challenges, which in turn create major obstacles to truly sustainable development. In Northern Ireland, the ongoing division of society along sectarian lines still prevents any 'normality' in political debate, from which any move towards sustainability could emerge. In the Republic of Ireland, the rapid exposure to global economics has placed intense strain on all levels of social identity and has had an impact on notions of kinship, community and nationalism, leading to tensions between majority society and minorities, both traditional and new (Fanning, 2002). Both parts of Ireland have significant lessons to learn from each other, not just in their localised experiences of sustainable development but also in their experience of issues of mediation, conflict management and social diversity. It is here that the uniqueness of Ireland as an island can begin to shape the approach to local sustainable development.

While most governance and citizenship benefits of sustainability have their prime focus at the local level, it does have some resonance in a cross-border context when we view the island as a single ecological unit. An enhanced recognition of the mutual island status of the two political units of Ireland may precipitate a fertile awareness of the issues that underlie sustainable development. Indeed, the idea of an island creates a very powerful metaphor of limits and can offer an important tool for sustainability analysis. The notion of an island also excites other passions, both cultural and psychological, with the physical 'boundedness' of an island capable of promoting 'bounded' identities and a resilience in the face of outside pressures (Hay, 2003). It has also been suggested that the island status of a community may also more readily form the 'bridging and bonding' ties that are basic to the development of social capital and sustainability (Kilpatrick and Falk, 2003).

This would therefore suggest that a gentle fostering of an 'island ethic' may offer new opportunities for the transition to sustainable development in Ireland. Despite some limited history of cooperation between the two jurisdictions on environmental management (Buick, 2002), this has not been extensive given the range of environmental issues such as water management, habitat conservation, tourism and pollution control that are directly affected by the border. This alone makes a strong case for cross-border cooperation and is accompanied by other arguments such as how the island forms a distinct economic region, with shared markets in many of the sectors that are critical to sustainable development. In governance terms, the emerging concept of European citizenship, such as trans-boundary participation rights in environmental decision-making (Macrory and Turner, 2002) are also forcing recognition of a common all-Ireland approach to participation and dialogue. However, somewhat ironically, LA21 is by design, focussed at the lowest level of government and in so doing can lead to a fetish of localism, as if it were disconnected from surrounding communities, albeit within a national and international policy framework. For the island of Ireland this has meant that local authorities in the Republic of Ireland and Northern Ireland, have looked away from each other, towards Dublin or Belfast and London respectively for guidance, cooperation and inspiration, rather than linking with other communities on the island. The development of an island ethic of sustainability is therefore a major project that faces significant ideological and constitutional barriers, but which could prove to be a critical stimulus to a broader range of objectives.

There are a number of initial steps that can be taken to begin to develop such an ethic and reinvigorate local sustainable development. One area, acknowledged by the UK government (DETR, 2000), but never having any official recognition by either Irish jurisdiction is the difficulty in communicating the objectives and benefits of sustainable development - while the term may have resonance amongst policy-makers, it is still largely unrecognised by the public. Although this may slowly improve as sustainability concepts are increasingly applied through public policy, the process could be speeded up and managed through coherent communication strategies in both parts of Ireland. This would not only provide local practitioners with advice on how best to articulate such issues to the public, but could be used to develop a coherent language more appropriate to the Irish context. Ideally this should also aim to forge a 'brand' of sustainability that incorporates the ideas of environmental citizenship and an island ethic. In the context of sustainability concepts being applied in an increased diversity of contexts, the coherence (and therefore understanding) of the concept may also be improved by the adoption of a marque for local sustainable development, to be used as an icon wherever any elements of sustainability is being delivered. Furthermore the very concept of environmental citizenship needs to be strengthened in the implementation of local sustainable development, by greater direct public engagement and more specific acknowledgment of, and direct support for social partners in the process. The cross-border dimension first needs to be promoted by the respective governments, and it is suggested that this could be stimulated by an All-Island Summit on Sustainable Development, the creation of an All-Island Local Sustainable Development Roundtable, and the development of cross-border networks of sustainability practitioners.

These initiatives clearly have to be accompanied by enhanced levels of government support and local capacity building within both Northern Ireland and the Republic of Ireland. In the North, there is a desperate need to articulate its own version of sustainability through a robust regional sustainable development strategy and more specific guidelines for the promotion of LAn21. In both jurisdictions there is a need to encourage increased level of dialogue between sustainability practitioners from each of the key sectors.

These proposals have been discussed in more detail elsewhere (Ellis et al., 2004) and it is important to realise that sustainable development is not a short-term project and requires significant changes in governance, partnership, decision-making and, indeed, everyday behaviour and cultural practices. The local sustainability agenda centres on embedding these principles into all aspects of local government and community development. Since governance and local development are at the heart of the project, society must look to local government to take the lead on these issues, albeit in a strengthened national policy context. Evidence suggests that local authorities are not currently adequately fulfilling this task and, generally speaking consider themselves 'too busy for that kind of stuff'. This could be turned around by one of a number of strategies, from enhanced resourcing to establishing sustainable development as a statutory objective. It is likely that such approaches will face difficulties in delivering cultural and behavioural change and this can only be achieved if local communities themselves can be invigorated to take such action. It has been argued here that a unique Irish sustainability ethic, based on citizenship and the ecological unity of the island should play a part in the development of such a future.

References

Agyeman, J. and Angus, B. (2003) 'The role of civic environmentalism in the pursuit of sustainable communities', *Journal of Environmental Planning and Management*, Vol. 46(3), pp. 345-364.

Audit Commission (1997) *It's a Small World: Local Government's Role as a Steward of the Environment*, London: Audit Commission.

Barry, J. (1996) 'Sustainability, political judgement and citizenship: connecting green politics and democracy', in D. Brian and M. de Geus (eds) *Democracy and Green Political Thought: Sustainability, Rights and Citizenship*, London: Routledge.

Beckerman, W. (1994) "Sustainable development': is it a useful concept?', *Environmental Values*, Vol. 3, pp. 191-209.

Brugman, J. (1997) 'Local authorities and Agenda 21', in F. Dodds (ed) *The Way Forward: Beyond Agenda 21*, London: Earthscan.

Buckingham-Hatfield, S. and Percy, S. (1999) *Constructing Environmental Agendas*, London: Routledge.

Buick, J. (2002) *Crossing the Border: a Regional Approach to Environmental Management, Report 2002.2*, Lund: International Institute for Industrial Environmental Economics.

Carley, M. and Chirstie, I. (1992) *Managing Sustainable Development*, London: Earthscan.

Church, C. and Young, S. (2001) 'The United Kingdom: mainstreaming, mutating or expiring?' in W.M. Lafferty (ed.) *Sustainable Communities in Europe*, London: Earthscan.

Comhar (2001) *Report to the Earth Council*, Dublin: Comhar.

Department for Environment, Food and Rural Affairs (UK) (DEFRA) (2003a) *Achieving a Better Quality of Life: Review of Progress Towards Sustainable Development*, Government Annual Report 2002, London: DEFRA.

Department for Regional Development (NI) (DRD) (2001) *Regional Development Strategy for Northern Ireland 2025*, Belfast: DRD, http://www.drdni.gov.uk/shapingourfuture.

Department of Arts, Heritage, Gaeltacht and the Islands (RoI) (DAHGI) (2002) *National Biodiveristy Plan*, Dublin: Stationery Office.

Department of Environment (UK) (1990) *This Common Inheritance*, London: HMSO.

Department of Health, Social Security and Public Safety (DHSSPS) (2002) *Investing in Health,* Belfast: DHSSPS, http://www.dhsspsni.gov.uk/publications/2002/investforhealth.html

Department of the Environment (NI) (2002) *Promoting Sustainable Living: a Discussion Paper on Proposals for a Sustainable Development Strategy for Northern Ireland*, Belfast: DoE (NI).

Department of the Environment (Republic of Ireland) (1995) *Local Authorities and Sustainable Development: Guidelines on Local Agenda 21*, Dublin: DoE.

Department of the Environment (Republic of Ireland) (1996) *Better Local Government – a Programme for Change*, Dublin: DoELG.

Department of the Environment (Republic of Ireland) (1997) *Sustainable Development: a Strategy for Ireland*, Dublin: Stationery Office.

Department of the Environment and Local Government (DoELG) (2000) *National Climate Change Strategy*, Dublin: Stationery Office.

Department of the Environment and Local Government (DoELG) (2001) *Towards Sustainable Local Communities: Guidelines on Local Agenda 21*, Dublin: DoELG.

Department of the Environment and Local Government (DoELG) (2002a) *Making Ireland's Development Sustainable*, Dublin: DoELG.

Department of the Environment and Local Government (DoELG) (2002b) *Preventing and Recycling Waste: Delivering Change*, Dublin: Stationery Office.

Department of the Environment and Local Government (DoELG) (2002c) *National Spatial Strategy for Ireland*, Dublin: Stationery Office.

Department of the Environment, Transport and the Regions (UK) (DETR) (1998) *Sustainable Local Communities for the 21st Century: Why and How to Prepare a Local Agenda 21 Strategy*, DETR, Cm4014, London: The Stationery Office.

Department of the Environment, Transport and the Regions (UK) (DETR) (1999a) *A Better Quality of Life: a Strategy for Sustainable Development*, Cm 4345, London: The Stationery Office.

Department of the Environment, Transport and the Regions (UK) (DETR) (1999b) *Quality of Life Counts*, London: The Stationery Office.

Department of the Environment, Transport and the Regions (UK) (DETR) (2000) *Towards a Language of Sustainable Development*, London: DETR.

Department of Social and Family Affairs (RoI) (DSFA) (2002) *National Anti-Poverty Strategy*, Dublin: Stationery Office.

Earth Summit Ireland (2002) *Telling It Like It Is: 10 Years of Unsustainable Development in Ireland*, Dublin: ESI.

Ellis, G., Motherway, B., Neill, W.J.V. and Hand, U. (2004) *Towards A Green Isle? Local Sustainable Development on the Island of Ireland*, Armagh: Centre for Cross Border Studies.

Environment and Heritage Service (EHS) (NI) (2002) *Northern Ireland Biodiversity Strategy*, Belfast: EHS.

Environment Protection Agency (1996) *State of the Environment 1996*, Wexford: EPA.

Fanning, B. (2002) *Racism and social change in the Republic of Ireland*,

Manchester: Manchester University Press.

Gilbert, R., Stevenson, D., Giradet, H. and Stren, S. (1996) *Making Cities Work: the Role of Local Authorities in an Urban Environment*, London: Earthscan.

Glass, S.M. (2002) 'Sustainability and local government', *Local Environment*, Vol. 7(1), pp. 97-102.

Hay, P. (2003) 'The Poetics of Island Place: articulating particularity', *Local Environment*, Vol. 8, No. 5, pp. 553-558.

Hom, L. (2002) 'The making of Local Agenda 21: an interview with Jeb Brugman', *Local Environment*, Vol. 7(3), pp. 251-256.

International Council for Local Environmental Initiatives (ICLEI) (2002) *LA21 Survey 2002*, Freiburg: ICLEI.

Isin, E. F. and Turner, B.S. (eds) (2002) *Handbook of Citizenship Studies*, London: Sage.

Jacobs, M. (1995) 'Sustainable development, capital substitution and economic humility: a response to Beckerman', *Environmental Values*, Vol. 4, pp. 57-68.

Kelly, R. (2000) 'Local Agenda 21 in the Mid-West region of Ireland', *Local Authority News*, Vol. 19(5), pp. 3-4.

Kilpatrick, S. & Falk, I. (2003) 'Learning in Agriculture: building social capital in island communities', *Local Environment*, Vol. 8 (5), pp. 501-512.

Knox, C. (1998) 'Local government in Northern Ireland: Emerging from the bearpit of sectarianism', *Local Government Studies*, Vol. 24(3), pp. 1-13.

Lafferty, W.M. (ed.) (2001) *Sustainable Communities in Europe*, London: Earthscan.

Lafferty, W.M. and Coenen, F. (2001) 'Conclusions and perspectives' in W.M. Lafferty (ed.) *Sustainable Communities in Europe*, London: Earthscan.

Lafferty, W.M. and Eckerberg, K. (1998b) 'The nature and purpose of "Local Agenda 21"', in W.M. Lafferty and K. Eckerberg (eds) *From the Earth Summit to Local Agenda 21*, London: Earthscan.

Lafferty, W.M. and Eckerberg, K. (eds) (1998a) *From the Earth Summit to Local Agenda 21*, London: Earthscan.

Local Government Management Board (LGMB) (1999) *Sustainable Local Communities*, London: LGMB.

Macrory, R. and Turner, S. (2002) 'Participatory rights, transboundary environmental governance and EC law', *Common Market Law Review*, Vol. 39, pp. 489-522.

Marvin, S. and Guy, S. (1997) 'Creating myths rather than sustainability: the transition fallacies for the new localism', *Local Environment*, Vol. 2(3), pp. 311-318.

Moser, P. (2001) 'Glorification, disillusionment or the way into the future? The significance of Local Agenda 21 processes for the needs of local sustainability', *Local Environment*, Vol. 6(4), pp. 453-467.

Mullally, G. (2001) 'Starting late: building institutional capacity on the reform of sub-national governance' in W.M. Lafferty (ed.) *Sustainable Communities in Europe*, London: Earthscan.

Myers, G. and Macnaghten, P. (1998) 'Rhetorics of environmental sustainability: commonplaces and places', *Environment and Planning A*, Vol. 30, pp. 333-353.

Northern Ireland Executive (2001) *Programme for Government 2001-2003*, Belfast: NI Executive.

Northern Ireland Executive (2003), *The Review of Public Administration in Northern Ireland*, Belfast: NI Executive.

North-South Ministerial Council (2000) *Joint Communiqué, Environment Sector, Interpoint*, Belfast, Wednesday 28 June 2000.

O'Riordan, T. (2002) 'Civic science and sustainability', in D. Warburton (ed.) *Community and Sustainable Development*, London: Earthscan.

O'Riordan, T. and Voisey, H. (1998) *The Transition to Sustainability: the Politics of Agenda 21*, London: Earthscan.

Otto-Zimmerman, K. (2002) 'Local Action 21: motto-mandate-movement in the post Johannesburg decade', *Local Environment*, Vol. 7(4), pp. 465-469.

Putnam, R. D. (2000) *Bowling alone - the collapse and revival of American community*, New York: Simon & Schuster.

Redclift, M. (1987) *Sustainable Development: Exploring the Contradictions*, London: Routledge.

Ryan, C. (2002) 'Local Agenda 21 in Action', in F. Convery and J. Feehan (eds) *Achievement and Challenge: Rio+10 and Ireland*, Dublin: The Environmental Institute, University College Dublin.

Selman, P. (1996) *Local Sustainability*, London: Paul Chapman Publishing.

Selman, P. and Parker, J. (1997) 'Citizenship, civicness and social capital in Local Agenda 21', *Local Environment*, Vol. 2(2), pp. 171-184.

Selman, P. and Parker, J. (1999) 'Tales of local sustainability', *Local Environment*, Vol. 4(1), pp. 47- 60.

Sharp, L. (2002) 'Public participation and policy: unpacking connections in one UK Local Agenda 21', *Local Environment*, Vol. 7(1), pp. 7-22.

Tuxworth, B. (2001) *Local Agenda 21: from the Margins to the Mainstream, and Out to Sea?* London: TCPA.

United Nations (1993) *Report of the United Nations Conference on Environment and Development, Rio de Janerio, 3-14 June 1992, Vol. 1: Resolutions Adopted by the Conference*, New York: United Nations.

Voisey, H. (1998) 'Local Agenda 21 in the UK', in T. O'Riordan and H. Voisey (eds) *The Transition to Sustainability: the Politics of Agenda 21 in Europe*, London: Earthscan.

Warburton, D. (2002) *Community and Sustainable Development*, London: Earthscan.

Young, S.C. (2000) 'Participation strategies and environmental politics: Local Agenda 21', in G. Stoker (ed.) *The New Politics of British Local Governance*, Basingstoke: Macmillan.

Chapter 10

Social and Ethnic Segregation and the Urban Agenda

Brendan Murtagh

Introduction

> To be off-balance and insecure in some primary or core identity can be, to use the terms currently in vogue, to occupy a position 'on the edge', an ill-defined 'borderland' from which the socially constructed nature of other identities wrestling with problems of 'Us' and 'Them' can be better appreciated (Neill, 2004, p. 1).

This chapter is concerned with dichotomous identities and how they produce 'edge' spaces in contemporary Belfast. Whilst it draws empirically on the peculiar insecurities of Northern society, it reflects on and aims to contribute to debates about multiculturalism and social segregation in modern Ireland. Irish urban growth has not been characterised by ethnic ghettoisation or residential segregation, which have been a feature of immigration in other advanced capitalist societies. However, Belfast still represents an important arena to explore 'the dark side of difference' (Sandercock, 2000) and the costs of concentrating ethnic-religious poverty in highly territorialized places. In particular, it argues that planners and urban managers have an important role to play in addressing the spatial working out of them and us politics, poverty and finding the real and figurative space to present alternative visions for multicultural urban society.

The chapter begins by setting ideas about ethnic and social segregation in context. It identifies the project of ethnic assimilation via residential integration as a distinctive strand in societies emerging from conflict and crucially in the context of Ireland, where multiculturalism is spatially expressed, especially in poor neighbourhoods. The chapter then links this with the emerging global concern for social exclusion and how this has reached mainstream urban policy debates in programmes such as Neighbourhood Renewal. It then examines how these narratives are brought together in contemporary Belfast and uses the EU URBAN II Community Initiative to explore the possibilities for policy in areas where race and poverty intersect to produce especially 'wicked problems'. Implications are drawn for urban policy in Ireland, especially in the fusion between equality and social justice, law and policy.

Identity, Citizenship and Spaces

> Planning, that is, the public regulation and 'production of space' is shown to serve as an instrument of social control. Like most other areas of public policy, it should thus be conceived as 'double-edged', being capable of both reform and control, emancipation and oppression (Yiftachel, 2000, p. 419).

The manipulative capacity of planning to exercise control through scarce land resources attracted particular attention in Israel and South Africa. Fenster (1996), for instance, identified the use of discriminatory processes and systems to deny Arabs significant land rights and entitlements in the Israeli state. She pointed out that discrimination could arise when ethnic groups are treated differently from the majority or when minorities are treated similarly in different situations. The former denies citizenship rights whilst the latter denies the uniqueness of ethnic group identities and lifestyles (Fenster, 1996, p. 415). For Safier (2001) the core of a planning agenda for the state should thus be explicitly centred on notions of civic identity and citizenship. The objective for urban planning is:

> To promote the idea of building a common civic identity and consciousness. This means the encouragement of a sense of living in, and identifying with, a city in which diversity of cultural communities, ways of life and distinctive practices are perceived as producing 'value added', that is a quality of city life that is greater than the sum of the individual constituent groups (Safier, 2001, p. 158).

Similarly, in South Africa, the project of assimilation has transformed the policy, if not the urban landscape of the new nation. Turok (1994) traced the transition from Apartheid to democracy and in particular, the effort invested in strategic planning to deal comprehensively with poor township infrastructure, the diversification of buffer zones used to isolate black areas and to construct appropriate forums to include community interests in the planning process.

But these alternatives are not confined to highly politicised territorial conflicts. For example, the Home Office response to race riots in the Northern British cities of Bradford, Burnley and Oldham in 2000 can be read, albeit loosely, within an assimilationist model (Amin, 2002). Here, bipolar politics, identities and education thwart the targets for a renewed initiative on community cohesion with a fairly selective rendition of British citizenship underpinning that policy drive:

> A civic identity, which serves to unite people and can express common goals and aspirations of the whole community, can have a powerful effect in shaping attitudes and behaviour. Shared values are essential to give people a common sense of belonging regardless of their race, cultural traditions or faiths (Home Office, 2001, p. 12).

Neo-liberal notions of assimilation also have had important appeal for activists of desegregation in the United States (Farley and Frey, 1994). Seitles claimed that "the devastating effects of residential racial discrimination on the quality of life for minority families and for our culture at large, represent the importance of initiating policies to integrate residential neighbourhoods. Without the efforts of integration, the negative effects of decades of bigoted housing policies will be exacerbated, therefore perpetuating the existence of segregation and racial division (Seitles, 1996, p. 17).

Amin (2002, p. 967) made the point that there is no guarantee that mixed neighbourhoods will be any less racist than segregated ones and that even "public spaces are not neutral servants of multicultural engagement". For this to happen, interaction in everyday life needs to be worked at through discursive negotiation that will ultimately build trust and reciprocation (Amin, 2002). Thomas (2000) further suggests that these are skills the urban planning profession does not posses in any great quantity or depth. The professions strong technical ideology, the interests of primary industries and powerful lobby groups, especially linked to the property sector, left little real room for a consideration of race and ethnicity, at least until the mid-1980s. This argument is developed by Bollens (2002, p. 37) who states that "planning should incorporate social-psychological aspects of community identity into its professional repertoire" by loosing some of its methodological 'certainty' and by seeking to understand racial diversity and the attachment that some ethnic groups have to land, and its ownership and control.

Social Exclusion, People and Place

A second, related narrative in this chapter is the concern with social exclusion and its interface with urban policy and land use planning. The crises of industrial restructuring and the decline of smokestack industries particularly since the 1970s caused governments to look to the private sector and new economy and less to dwindling state resources to lead change. Market planning, levering in private development money and stimulating the property economy underpinned the ideology of the New Right and its manifestation in Thatcherism and Reaganism in the 1980s. The urban renaissance of waterfronts, docklands and town centres reflected the partial success of these initiatives but not everyone or every area benefited from new expensive consumption arenas and the jobs that required more advanced skills in finance, communications or informatics. The rediscovery of poverty in politics, European policy and in spatial analysis drew attention to the limits of the market and stimulated new theoretical and policy debates on social exclusion and neighbourhood crises (Soja, 2000). Three strands can be identified in the literature plotting the turn within urban policy toward social objectives, communities and new forms of urban governance. First, the priority given to physical and property development represented only a partial analysis of the complex and interrelated nature of urban change. Ginsburg (1999) pointed out that the social fabric of declining neighbourhoods became manifest as a wider crisis with crime, low educational attainment, ill-health and fatalism sparking inner city riots and racial alienation. Second, and linked to this was what Duffy and Hutchinson (1997) referred to as the 'turn to community' in the way in which urban managers developed local plans and invested in regeneration projects. Sidelined in the 1980s, 'people' became the target of strategies to develop solidarity, 'social capital' and indigenous capacity in order to take a leading role in urban programme delivery. In the United States, Community Empowerment Zones and in Britain, Neighbourhood Renewal Areas became the focus for new investment in community skills and structures (Healey, 2003). Recently, the spatial target had shifted away from high growth commercial sites in central business districts and waterfronts and toward areas of disadvantage. New methods for mapping multiple deprivations and consulting communities emerged in this period of policy shift. Feeding from this is the third main strand, which involved the development of new governance arrangements to include a range of stakeholders involved in local decision-making and strategy management (Healey, 2002). Area based partnerships involving communities, the private sector and government took responsibility for the development of local needs analysis, comprehensive plans and project implementation (Ginsburg, 1999). Increasingly, European urban and regional policy has reflected these broad shifts. Publicly-led investment in tightly defined communities, using local people to determine priorities, integrated strategy development and an explicit concern for ethnic and social exclusion became consistent features of guidance and specific programmes in regional policy (Williams, 1999). The European Commission launched the URBAN Community Initiative in 1994 as a way of developing a common European understanding of contemporary urban change and local development possibilities. The next section looks at how this shifting policy agenda has played out in the distinctive conditions of Belfast and what the wider implications might be for programme implementation across the country.

Ethnic and Social Division in Belfast

Attachment to land and ethno-religious segregation has been a feature of Belfast for at least five generations (Boal, 1996). Segregated housing areas, once initiated are reproduced by patterns of residential mobility and crucially, violence (Boal, 1996). This section attempts to summarise some of the planning effects of ethnic religious segregation in the city and policy responses especially in the most intense periods of conflict. In short, it asserts that the material effects of territory make a compelling case for urban policy to intervene and for planners to move beyond technocracy and appreciate the wider nuances of ethnic-religious space.

Recent patterns of industrial restructuring in the Belfast Metropolitan Area (BMA) has produced differential effects especially on inner-city communities traditionally reliant on jobs in heavy engineering and manufacturing (Figure 10.1). The new geography of work has created service sector jobs in different production sites in the city centre, Laganside and suburban service hubs. Yet

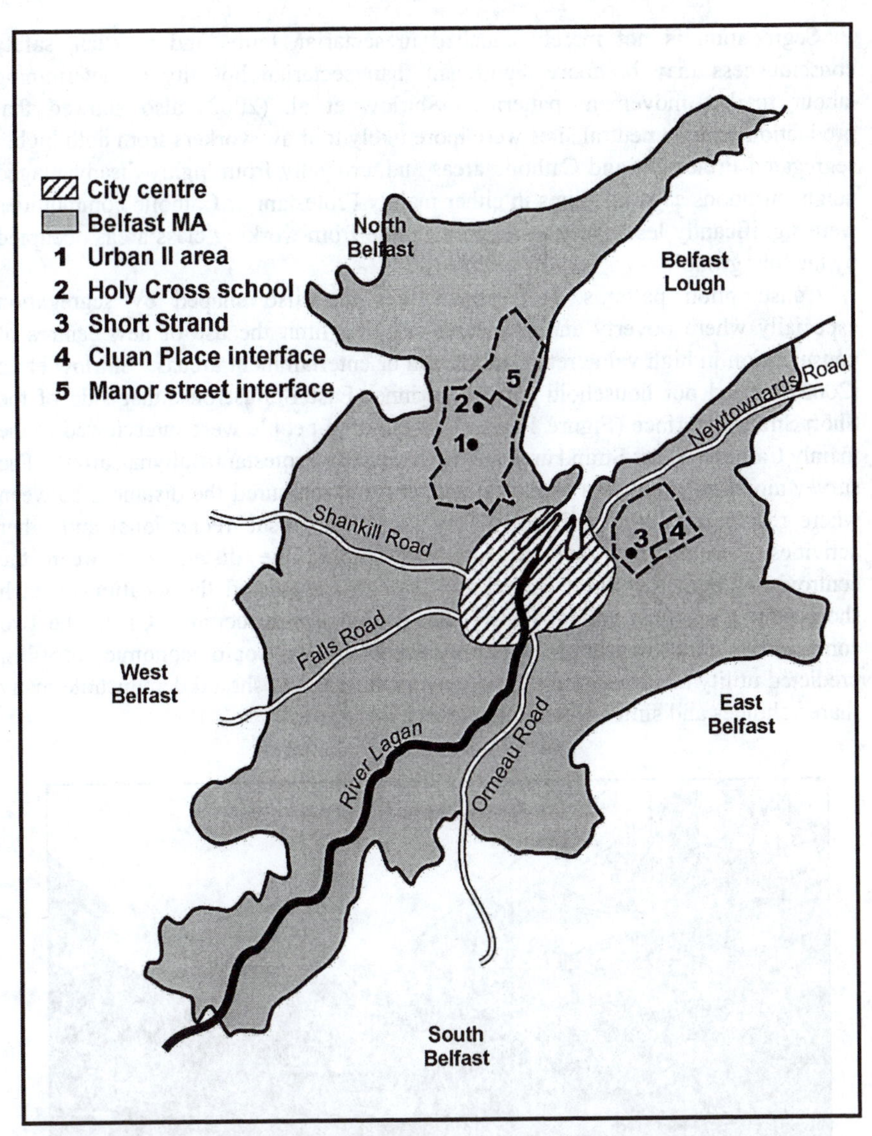

Figure 10.1 Key locations in the Belfast Metropolitan Area

segregation continues to assert its influence on where people work and how safe they feel both going to and being in their workplace. Shirlow et al. (2002) have looked at the effects of territory on 13 separate private sector businesses located in different parts of the city by measuring the spatial distribution of their workforce by religion and social class areas. The study covered 10,418 employees or about 4 per cent of employees in the BMA and found that, when analysing the religious mix of the workforce. The location of employment, in sectarian terms, is more important than the frictional effects of distance.

Segregation is not merely couched in sectarian terms and as such, safety consciousness may be more significant than sectarian hostility in determining labour market movement patterns. Shirlow et al. (2002) also showed that production units in neutral sites were more likely to draw workers from both highly segregated Protestant and Catholic areas and, crucially from highly disadvantaged neighbourhoods as well. Sites in either mainly Protestant or Catholic communities were significantly less likely to draw in people from working class areas occupied by the 'out group'.

Consumption patterns and opportunities are also shaped by segregation especially where poverty and low car ownership limit the use of new centres of consumption in high value retail, recreation or entertainment arenas. Shirlow et al. (2002) carried out household surveys in inner East Belfast on either side of the Short Strand interface (Figure 10.2). Two hundred people were interviewed in the mainly Catholic Short Strand area and in the mainly Protestant Ballymacarrett. The survey aimed to produce a statistical analysis that measured the distances between where respondents live and where they shop and pursue recreational and other activities. Interactions are measured as straight-line distances between the centroids of both the Short Strand and Ballymacarrett and the location of each shop, pub, club, post office, library, health and leisure centre. Given the two communities' close spatial proximity and similar socio-economic profiles, predicted utility maximisation should be very similar and should demonstrate many shared choices and similar interaction levels.

Figure 10.2 Interface at Cluan Place, East Belfast

A statistical measurement of dissimilarity in consumer choice can be drawn out by a simple coefficient of determination (r2). Table 10.1 lists highly insignificant relationships for all seven services and facilities, where the r2 value is closer to 0.0 than +/-1.0. This indicates that there is virtually no relationship or association between where consumers from the Short Strand travel and where consumers from Ballymacarrett travel for their respective services and facilities. The highest values of association are for consumers using health centres (r2=0.312) and leisure centres (r2=0.262). For health centres, the primary reason is the sharing of local GP and health centre facilities. Similarly, the leisure centre is located on a non-residential major arterial route that residents from the Short Strand feel they can access safely and quickly.

People in interface communities in Belfast are also twice as likely to be unemployed and on low incomes than the rest of the city's population (Murtagh, 2002). Those with most power in the housing system choose to leave and associated with this 'exiting' is a process of residualisation whereby communities occupying marginal, dangerous or contested territory are increasingly characterised by high rates of social deprivation and poverty. Added to this are the daily experiences of people living in an area affected by violence, pervasive

Table 10.1 Community interaction and mobility in East Belfast

Service	*r*	*r2*	*Total trip distance (km) Short Strand*	*Total trip distance (km) Ballymacarrett*	*Mean trip distance (km) Short Strand*	*Mean trip distance (km) Ballymacarrett*
All services	0.202	0.041	1364.72	1111.69	1.51	1.25
Daily shopping	0.219	0.048	253.72	129.94	2.54	1.30
Leisure centres	0.512	0.262	145.41	76.57	2.79	1.22
Health centres	0.559	0.312	35.42	57.20	0.43	0.71
Public houses	0.361	0.130	50.31	80.60	0.65	1.26
Post office	0.200	0.040	98.66	58.32	1.10	0.75
Library	0.226	0.051	149.94	58.57	3.57	1.27
Social Security	0.390	0.152	120.12	70.79	2.50	1.77

Source: Shirlow et al. (2002, p. 83)

fear and the threat of attack. For example, a joint study by the Housing Executive and the Eastern Health and Social Services Board found that the residents of the Protestant enclave of Suffolk in West Belfast experienced higher rates of long standing illness, attendance at GPs and stress than the rest of Northern Ireland population (NIHE and EHSSB, 1995). Deaths and injuries as a consequence of 'The Troubles' were also highest in areas where ethno-religious space was most contested. The study also showed that one-third of the victims of politically motivated violence were killed within 250 metres of an interface and 70 per cent of deaths occurred within 500 metres of a peaceline. Moreover, it was concluded that the negative image that brutal lines of division, physical dereliction and poverty projected to a wider and, in particular, international audience could be a major obstacle to inward investment and tourism (Neill, 1999).

State Responses

In the last thirty years, the Northern Ireland crisis has been met with a multiple response involving militarisation, social reform and legislative guarantees for the Catholic minority, especially in the labour market. An explicit assimilationist project was supported with the establishment of the Community Relations Council, the promotion of integrated education and an attempt to restore imbalances in the economic conditions of Catholics and Protestants (Shirlow, 2001). The formation and maintenance of even a loose form of material citizenship offered an opportunity to relax atavistic identity constructions. Economic restructuring and equality of opportunity guaranteed new mobility for a rising Catholic middle class who saw less investment in the Nationalist project and could identify the benefits of apolitical lifestyles and aspirations (Porter, 1998). Planning played an important role in modernising the city, providing new consumption opportunities for this emerging disposable income and projecting positive place imagery to global investors and tourists. This class realignment was reflected in tenure restructuring with a residualised public sector housing stock containing comparatively even numbers of Catholics and Protestants and new suburban, riverside and gentrified housing spaces responding to social, spatial and tenurial mobility within the city (Neill, 1999). New spatial cleavages were overlain on the traditional territorial landscape but old enmities were most viciously contested on the sink estates of North and West Belfast where 25 peacelines separated Catholics and Protestants abandoned by economic change and largely untouched by progressive social need or equity policy (Ellis, 2001). Assimilation is checked by plurality and its deeply socio-political and spatial character. The modernising (or post modernising) of contemporary Belfast left little space for any serious consideration of the planning specifics of territoriality and urban segregation. Security force management not urban policy was the primary response to 'sites of resistance' and the technocratic and colour-blind response of the planning system helped to insulate it from any accusation of discrimination or bias (Bollens, 1999).

A significant shift in this policy stance was created by both global and locally specific factors. It was pointed out earlier in the chapter that the failure of liberation economics and property market planning to trickle down to the most disadvantaged communities refocused attention on communities, social objectives and new governance arrangements (Healey, 2002). In Britain, this was reflected in an explicit policy focus of the Blair administration on social exclusion and neighbourhood renewal. This was concordant with European urban policy priorities for longer-term programmes aimed at restructuring local economies and integrating marginal groups into the labour market (Murtagh and McKay, 2003).

Locally, these shifts were given a distinctive flavour by the signing of the Good Friday Agreement and the promotion of equality at the heart of Government policy (Ellis, 2001). The Northern Ireland Act 1998, which provided the legal interpretation of the Agreement, set down formidable equality duties on all public bodies in the region. The Act established a Northern Ireland Human Rights Commission and introduced a statutory duty on public bodies to have:

> Due regard to the need to promote equality of opportunity:
> (a) between persons of different religious belief, political opinion, racial group, age, marital status or sexual orientation;
> (b) between men and women generally;
> (c) between persons with a disability and persons without;
> (d) between persons with dependents and persons without (Section 75(1)).

In carrying out their functions, public authorities must also "have regard to the desirability of promoting good relations between persons of different religious belief, political opinion or racial group" (Section 75(2)). Government Departments had to prepare Equality Schemes that showed how these objectives would be met through current policies and Equality Impact Assessments were introduced to proof all new programmes against the needs of the groups identified in Section 75(1).

Coupled with the legislative emphasis on equality was a concern for Targeting Social need and Promoting Social Inclusion. A new spatial map of deprivation, the Noble Index,[1] helped to direct urban and rural development programmes and in particular, the Department for Social Development's Neighbourhood Renewal Strategy, *People and Place* (DSD, 2003). *People and Place* identified segregation as a significant component in understanding the multi-layered nature of urban disadvantage and the need to tackle the blighting effects of interfaces in wider regeneration programmes. The document pointed to innovation in the EU URBAN II Community Initiative Programme (CIP) 2000-2006 for North Belfast as evidence of the Government's commitment to this task. The next part of the chapter looks at the formulation of the Initiative and in particular its attempts to grapple with the spatial legacy of conflict, the cross-community structures needed to deliver the programme and the skills required to sustain neighbourhood regeneration in the longer term.

Case Study: URBAN II, North Belfast

Setting the Context

The introduction of the URBAN II CIP reflected the broader European concern for the future of the city region and drew on pilot projects, the experience of Integrated Development Operations and the impact of URBAN I (Chorianopoulos, 2002). The Programme has two objectives:

- To promote the formulation and implementation of particularly innovative strategies for sustainable economic and social regeneration of small and medium sized towns and cities or of distressed urban neighbourhoods in larger cities;
- To enhance and exchange knowledge and experience in relation to sustainable urban regeneration and development in the Community.

In Belfast, the Programme was targeted on inner North Belfast (31,000 people) and is designed to invest €17.10m between 2000 and 2006. The selection of the area followed a period of extensive consultation, which highlighted the particularly intense nature of local problems. First, the area is stratified by six peacelines and by a long history of fear, sectarian deaths and internecine conflict (Figure 10.3).

Figure 10.3 Interface in Manor Street, URBAN II area

Second, North Belfast has suffered disproportionately from the effects of deindustrialisation, redevelopment and depopulation. The lost jobs in primary industries were not replaced by new employment in growth sectors of the economy. Low skills, educational attainment and even basic literacy and numeracy skills reflected the areas wider decline. The Noble Index showed that the most disadvantaged cluster of wards in Northern Ireland were concentrated in North Belfast (DSD, 2001). A significant component of the area identification process was that it was supported by the cross-party Social Development Committee and by the Democratic Unionist Party (DUP) Minister for Social Development, Nigel Dodds, before the subsequent suspension of Assembly business.

Table 10.2 illustrates the particular problems and development opportunities in the North Belfast URBAN II area. The SWOT analysis sets the context for the URBAN II Strategy and highlights the deeply structural nature of the areas economic demise:

- In particular, it identifies the need to restructure redundant and blighted land and buildings and regenerate the property market especially in a way that meets community needs and priorities. Key development opportunities might be defined given the nature of the spatial problems and the sheer quantity of derelict sites in the designated area.
- There is also a need to restructure the labour market and promote realistic job chances for local people. This is especially important in any attempt to integrate young people, the long-term unemployed and men and women into the labour market.
- There is also the need to restructure the community infrastructure of the area, reduce tension and fear and facilitate a process of bottom up renewal in North Belfast.
- Finally, there needed to be a greater sense of corporate and integrated planning in the area to maximise resources as well as specific technical skills to allow the local community and potential beneficiaries to engage in the programme effectively.

Strategy Delivery

Following this analysis the strategy attempted to develop an integrated approach captured by the overall mission statement:

> To regenerate inner North Belfast into a vibrant, safe and viable urban community for its people, its environment and its economy.

Specifically, it developed operational objectives, which reflected the economic, environmental and social problems facing the North Belfast area, namely "to develop the physical and social resources of the area; to develop the resources of

Table 10.2 A SWOT analysis of North Belfast

Strengths	*Weaknesses*
• Land capacity and development opportunities; • Infrastructure and arterial road links; • Proximity to M3 and West link; • Views and natural amenity resources; • Programmes at a community level; • Industrial zones located in and around the area; • Improved housing stock; • Specific programmes and consultation process emerging in areas such as housing; • North Belfast Partnership and cross sectoral community management of local problems.	• Deprivation and various forms of social exclusion; • High and persistent rates of local unemployment; • The blighting effects of sectarian interfaces; • Poor condition of key arterial routes and access points; • Fractured labour markets; • Poor rate of job creation; • Large amount of derelict land and property; • Poor educational attainment and training provision; • Low skills base among economically active population; • Segregation, fear and low cross-community trust; • Weak community infrastructure and organisational capacity; • Lack of a coordinated strategic focus to the development of the area; • Limited community skills in mainstream regeneration planning; • Poor community health.
Opportunities	*Threats*
• New strategic sites coming on stream such as Crumlin Road Courthouse, Jail and Army barracks; • Possibilities at local level opened by new political space; • Cross community management in infancy but happening; • Labour market supply within the area; • North Belfast high of the strategic priorities of a number of development agencies; • Economic and physical	• Decline in manufacturing and high rates of LTU among both men and women; • Demographic decline and imbalance between the religions; • Further ethno-religious residential segregation; • Violence and dysfunctional community relations that could limit the prospects of cross community negotiations; • Area becomes too inhospitable to external investors; • Alienation of young people, drug

development opportunities linked to the strategic roads network; • Experiences developed by communities in URBAN I could be transferable to the designated area; • Developing the private sector housing market linked to the production of a safe and attractive environment.	dependency and teenage pregnancy; • Community alienation from planning and local regeneration processes; • Increases in sectarian crime, intimidation and community fear.

Source: DSD, 2001, p. 59

local people especially to gain access to lasting employment opportunities; and to develop and apply the technical competencies required ensuring that the Programme is implemented effectively and efficiently" (DSD, 2001, p. 40).

The Programme is being delivered by the North Belfast Partnership, which has drawn together representatives from the statutory, private and community sectors to manage the wider regeneration of this part of the city. The Partnership had struggled to hold together Protestant and Catholic interests on one Board, especially given the intense violence around the Holy Cross dispute and the effects of the traditional marching season. The Holy Cross dispute involved a street protest by Protestant residents of the Glenbryn area against Catholics from Ardoyne who walked through the housing estate to bring children to a girl's primary school. The issue reflected a wider contestation of territory in North Belfast that is also manifest in Orange Order parades, especially in July, when the Order celebrates the victory of the Protestant armies at the Battle of the Boyne in 1690. But, intra as well as inter community rivalries including paramilitarism and organised crime, added further to the contested nature of North Belfast. Allocating scarce financial resources in this environment is a delicate 'no-win' task and the very fact that a Programme Management Executive has been established to deliver the Measures and meet European Union Decommitment targets on expenditure is an achievement in itself. Progress has been slow in dealing with interfaces and contested sites but the URBAN II Manager points out that there is no alternative to the slow iterative process of building confidence, some mutual trust and ultimately viable project ideas. As Amin notes "there is no formula here other than the engineering of endless tasks of interaction between adversaries or provision for individuals to broaden their horizons, because any intervention needs to work through, and is only meaningful in a situated dynamic" (Amin, 2002, p. 969).

A number of flagship projects are being brought forward for implementation on or near an interface underpinned by painstaking work with communities on both sides. An obstacle to the more rapid delivery of the initiative is the differential capacities of the Protestant and Catholic communities in the area. Protestants appear to have less experience, people and skills in local development whilst Catholic communities have a stronger track record in negotiating their way through the grant regimes of Europe as well as mainstream regeneration programmes (Shirlow and Murtagh, 2004). Building 'binding' and 'bridging' social capital is a priority for the Initiative but what is less clear is what skills, at what level and in what sectors the investment should be made. The need for structured support in training, education and professional development in issues such as the social economy, understanding financial systems and project management is pressing if the Initiative is not to pander to sectarian interests concerned with small scale localised projects rather than the scale of project needed to restructure the area's economy and environment. It would be wrong to overstate the impact of URBAN II on North Belfast's stubborn social and ethnic problems, but it is a start. Fundamentally, it marks a departure in policy thinking about the link between identity, segregation and urban renewal. Space as a problem and opportunity is being redefined in a way that makes a clear connection between equality and social need in actual delivery of programmes. Despite the politics of territory, the North Belfast Partnership has managed to establish and maintain a robust governance structure and spend money on a cross-sector and cross-community basis.

Policy Impact

The emerging Neighbourhood Renewal Strategy in Northern Ireland draws on this experience in a more mainstream effort to recognise the reality of planning in a deeply divided city (DSD, 2003). In particular, it has drawn on the approach to spatial analysis and used a systematic GIS based methodology, including the Noble Index, to determine areas that experience poorest conditions. The approach is also based on an integrated strategy with four policy objectives that address economic, physical, social and community themes. The approach to Neighbourhood Renewal is also based on partnership delivery and drawing on the early experiences of URBAN II and in particular, on the way in which structures that located decision making as close to the target community as possible have value in developing sustainable regeneration outcomes. However, an important struggle within the early delivery of URBAN II has been to integrate the programmes of Government Departments and agencies as well as private sector resources with those of the public sector. Belfast is a city with a remarkably complex set of governance arrangements, which are in part a legacy of Direct Rule and the weak role of local government (Parkinson, 2004). A range of Departments, local Councils, agencies and less formal partnership bodies have responsibility for planning, housing, urban regeneration and community development. Achieving any sort of integration and crucially leadership, in the field of citywide renewal has been compromised by fractured governance arrangements, overlapping functions and even competition between agencies delivering redevelopment programmes:

> Governance and decision making is fractured and inefficient. There are a plethora of strategies for different parts of Belfast. But there is much less indication that these are capable of actually being delivered in a joined up manner (Parkinson, 2004, p. 7).

There is a need to see URBAN II in the context of strategic, linked, corporate planning for the city, which integrates deprived areas into a clearly stated vision for Belfast. For this to happen, the pace of political progress, the stability of the devolved administration and ultimately effective metropolitan-scale local government needs to be put in place and sustained over time.

Conclusion

In Belfast, space has and continues to be manipulated as a cultural resource, electoral platform and safe haven for paramilitarism. Increasingly, it is renegotiated in the maintenance of petty power structures and criminality with schooling, housing and public transport quickly sectarianised as the political and paramilitary extremes redefine themselves in the new dispensation. Post-modern Belfast has segmented space and publicly speaks to the new realms of consumption in the city centre and the riverfront. Presumably, Belfast's failure to get short-listed as a City of Culture in 2008 can, in part, be explained by the panellist's ability to see beyond this rendition to view the places stratified by poverty and fear. This 'other' city can no longer afford to be conceived as a security issue and if equality and social need are fused properly a spatial consciousness might emerge between the oppressed living on either side of the peacelines. It could offer a normative framework for planners and urban managers to contribute positively to the emerging political and policy landscape. The tentative signs in initiatives such as URBAN II are encouraging but the lack of skills limit the impact of this agenda. Professional ideology rooted in a late capitalist urban world says little about the need to understand the spatial constraints of ethnic segregation. New competencies are required to understand places saturated with all sorts of real and symbolic meaning, conflict mediation, cross-community participatory practices, dispute resolution and governance. As race, ethnicity and poverty become increasingly important features of global urban life, this task has application beyond the particular confines of Northern Ireland.

The use of space to discriminate and marginalize as well as to target inequalities are placed in sharp relief in Belfast, Israel and South Africa. Socio-cultural discrimination and injustices are less visible in the landscape of urban Ireland but policy systems have produced differentiated social outcomes. McGuirk (2000) has identified the closed and corporatist systems that have characterised the development and delivery of the Custom House Docks scheme. These marginalized or excluded community interests and effectively by-passed local authority structures in regenerating the city's riverfront. Similarly, Moore (1999) highlighted the way in which urban schemes have disproportionately benefited commercial and property interests rather than disadvantaged communities in Greater Dublin.

Traditionally, Irish urban policy has relied on incentivising the market to stimulate regeneration well before Thatcherism or Reaganism celebrated the concept. Fiscal and grant regimes have levered in the private sector in the hope of achieving wider planning gain in local communities. Sustained economic growth and the re-distributive potential of economic surpluses have been identified as a political and policy issue in the National Anti Poverty Strategy (NAPS). Spatial poverty and exclusion is a growing priority as many of those who have not benefited from the growth economy are increasingly corralled in marginal inner and outer city housing estates. Ireland has a wealth of practice to draw upon, especially in terms of community-led development, partnership working and the social economy. Some of these have worked their way into parts of the Integrated Area Programme approach and the URBAN II Community Initiative Programme in Ballyfermot in North Dublin. But, debt financing areas of multiple and complex deprivation and not seeking short-term market-led options needs stronger mainstream resources and policy support. NAPS provides one potentially overarching framework to develop a clearer community-led regeneration approach to areas of high disadvantage. As Ireland becomes a more culturally diverse society avoiding the spatial nadir of ethnic segregation and its correlation with poverty will be the real test of the value of equality and anti-poverty strategies in the state.

Note

[1] The Noble Index is a synthetic index combining 52 variables at Ward and Enumeration District level to describe socio-economic deprivation in the relevant area. The 52 variables are grouped together into 7 *domains*, which are income; employment; health and disability; education skills and training; access services; social environment; and housing stress.

References

Amin, A. (2002) 'Ethnicity and the multicultural city: living with diversity', *Environment and Planning A*, Vol. 34(1), pp. 959-980.

Boal, F. (1996) 'Integration and division: sharing and segregation in Belfast', *Planning Practice and Research*, Vol. 11(2), pp. 151-158.

Bollens, S. (1999) *Urban Peace-Building in Divided Societies: Belfast and Johannesburg*, Boulder, CO: Westview Press.

Bollens, S. (2002) 'Urban planning and intergroup conflict: confronting a fractured public interest', *Journal of the American Planning Association*, Vol. 68(1), pp. 22-42.

Chorianopoulos, I. (2002) 'Urban restructuring and governance: North-South differences in Europe and the EU URBAN Initiative', *Urban Studies*, Vol. 39(3), pp. 705-726.

DSD (Department for Social Development) (2001) *Operational Programme for the URBAN II Community Initiative Programme 2000-2006*, Belfast: DSD.

DSD (Department for Social Development) (2003) *People and Place, Northern Ireland Strategy for Neighbourhood Renewal*, Belfast: DSD.

Duffy, K. & Hutchinson, J. (1997) 'Urban policy and the turn to community', *Town Planning Review*, Vol. 86(3), pp. 347-362.

Ellis, G. (2001) 'Social exclusion, equality and the Good Friday Agreement: the implications for land use planning', *Policy and Politics*, Vol. 29(4), pp. 393-411.

Farley, R and Frey, W. (1994) 'Changes in the segregation of whites from blacks during the 1980s: small steps toward a more integrated society', *American Sociological Review*, Vol. 59, pp. 23-45.

Fenster, T. (1996) 'Ethnicity and citizen identity in planning and development for minority groups', *Political Geography*, Vol. 15, pp. 405-418.

Ginsburg, N. (1999) 'Putting the social into urban regeneration policy', *Local Economy*, Vol. 13(5), pp. 55-71.

Healey, P. (2002) 'On creating the 'city' as a collective resource', *Urban Studies*, Vol. 39(10), pp. 1777-1792.

Healey, P. (2003) 'Collaborative planning in perspective', *Planning Theory*, Vol. 2(2), pp. 101-123.

Home Office (2001) *Building Cohesive Communities: A Report of the Ministerial Group on Public Order and Community Cohesion*, London: Home Office.

McGuirk, P. (2000) 'Power and policy networks in urban governance: local government and property regeneration in Dublin', *Urban Studies*, Vol. 37(4), pp. 651-672.

Moore, N. (1999) 'Rejuvenating docklands: the Irish context', *Irish Geography*, Vol. 32(4), pp 651-72.

Murtagh, B. (2002) *The Politics of Territory*, London: Palgrave.

Murtagh, B. and McKay, S. (2003) 'Evaluating the social effects of the EU URBAN Community Initiative Programme', *European Planning Studies*, Vol. 11(2), pp. 193-211.

Neill, W. (1999) 'Whose city? Can a place vision for Belfast avoid the issue of identity', *European Planning Studies*, Vol. 7(3), pp. 269-281.

Neill, W. (2004) *Urban Policy and Cultural Identity*, London: Routledge.

NIHE and EHSSB (Northern Ireland Housing Executive and the Eastern Health and Social Services Board) (1995) *Health and Housing*, Belfast: NIHE.

Parkinson, M.. (2004) *A Competitive City*, Belfast: Belfast City Council.

Porter, N. (1998) *Rethinking Unionism: An Alternative Vision for Northern Ireland*, Belfast: Blackstaff.

Safier, M. (2001) 'The struggle for Jerusalem arena of nationalist conflict or crucible of cosmopolitan coexistence?', *City*, Vol. 5(2), pp. 135-168.

Sandercock, L. (2000) 'When strangers become neighbours', *Planning Theory and Practice*, Vol. 1(1), pp. 13-30.

Seitles, M. (1996) 'The perpetualisation of residential racial segregation in America: historical discrimination, modern forms of exclusion and inclusionary remedies', *Journal of Land Use and Environmental Law*, Vol. 14(1), pp. 1-30.

Shirlow, P. (2001) 'Devolution in Northern Ireland/ Ulster/ the North/ Six counties: Delete as appropriate', *Regional Studies*, Vol. 35(8), pp. 743-752.

Shirlow, P., Murtagh, B., Mesev, V. & McMullan, A. (2002) *Measuring and Visualising Labour Market and Community Segregation*, Coleraine: University of Ulster.

Shirlow, P. and Murtagh, B. (2004) 'Capacity-building, representation and intra community conflict', *Urban Studies*, Vol. 41(1), pp. 57-70.

Soja, E. (2000) *Postmetropolis: Critical Studies of Cities and Regions*, London: Blackwell.

Thomas, H. (2000) *Race and Planning: The UK Experience*, London: UCL Press.

Turok, I. (1994) 'Urban planning in the transition from Apartheid', *Town Planning Review*, Vol. 65(4), pp. 243-259.

Williams, R. (1999) 'European Union: social cohesion in town planning', in Greed, C. (ed.) *Social Town Planning*, London: Routledge, pp.127-143.

Yiftachel, O. (2000) 'Social control, urban planning and ethno-class relations: Mizrahi Jews in Israel's 'Development Towns'', *International Journal of Urban and Regional Research*, Vol. 24(2), pp. 418-437.

Chapter 11

The State and Civil Society in Urban Regeneration: Negotiating Sustainable Participation in Belfast and Dublin

Jenny Muir

Introduction

Developing some of the ideas discussed in the previous chapter, this chapter considers the results of case study research into two urban regeneration programmes: the North Belfast Housing Strategy in Northern Ireland, and the Ballymun regeneration initiative in Dublin, Republic of Ireland. The chapter begins with a review of the promotion of public participation in urban regeneration programmes and in sustainable development policies in both Irish jurisdictions. A theoretical discussion follows, highlighting that the interaction between state agencies and civil society in urban regeneration is mediated through participation structures, and a definition of sustainable participation is proposed. Participation is then presented as a hegemonic project within urban regeneration programmes, operationalised through partnership structures. A framework is put forward to assist the analysis of participation processes as a site of interaction between the state and civil society, in order to assess the effectiveness of participation processes in urban regeneration programmes. The case study research findings are then presented, depicting the gap between the ideology of participation and its complex, uneven reality. Finally, some conclusions are drawn about the nature of participation in the case study areas and some other more general issues.

Urban Regeneration, Participation and Sustainability

Public Participation in Irish Urban Regeneration Programmes

The roots of public participation in Irish urban regeneration programmes (North and South) lie in two factors that caused changes in the structures of social administration in western democratic societies: the growth of protest movements during the mid to late 1960s, and the evolution of fiscal crisis during the early to mid 1970s. Governments sought to incorporate public protest, and to gain legitimation for projects that often required cuts in public expenditure, through a move away from relying solely on representative democratic structures towards the inclusion of participatory forums that allowed citizens to contribute directly to decision-making. Previously, mass participation in any form other than voting had been regarded as undemocratic by political theorists (Pateman, 1970).

Changes took place more quickly in Britain than in the Irish jurisdictions, with public participation incorporated into British urban regeneration programmes through partnership structures from the mid-1970s onwards (Imrie & Raco, 2003). Northern Ireland was affected in a particular way by the protest movements of the 1960s. The strengthening of agitation for equal rights for the minority Catholic community led to the resurgence of older community divisions through severe inter-community violence (Bardon, 1992). A subsequent restructuring of public administration in response to the McCrory Report of 1970 removed many services from the control of locally elected representatives (McCready, 2001). Northern Ireland was protected from the worst affects of the United Kingdom (UK)'s fiscal crisis of the 1970s because the priority was to maintain public order rather than to cut public expenditure. The question of the legitimacy of the British state within the province did not focus on economic management but more fundamentally the right to govern.

Within this context, Northern Ireland had fewer urban regeneration programmes than the rest of the UK. Several *ad hoc* local partnerships were initiated before the paramilitary ceasefires of 1994 (details in Greer, 2001). After the 1994 ceasefires, a network of District Partnerships was established to implement the European Union (EU) PEACE I Programme (Hughes et al., 1998; Murtagh, 2001). A modified form of District Partnerships known as Local Strategy Partnerships is administering the subsequent EU Peace II programme. Northern Ireland's recent Urban Regeneration Strategy targets the most deprived areas and emphasises the importance of partnership with local people (DSD, 2003), an approach similar to England's *National Strategy for Neighbourhood Renewal* (Social Exclusion Unit, 2001). The province's *Targeting Social Need* framework is also being revised, with the intention of developing an Anti-Poverty Strategy.

Practices of social administration in the Republic of Ireland were not affected by the protest movements of the 1960s, despite the existence of some small-scale campaigns. Fiscal crisis also came later, in the early 1980s (Adshead & Quinn, 1998). The growth of public participation was evident from the mid-1980s onwards, starting with the first and second EU anti-poverty programmes and continuing after 1988 when the reform of the EU Structural Funds included a commitment to the principle of partnership. The Irish Republic's response to fiscal crisis was to co-operate with rather than to confront organised labour. In 1987 the corporatist social partnership arrangement was established, consisting of agreements between public, private and voluntary 'pillars' (Walsh et al., 1998). Public participation in urban regeneration programmes also drew upon earlier traditions of voluntary action, such as the involvement of the Catholic Church in the delivery of social services, and community-based economic development projects in rural areas (Curtin and Varley, 1995).

Area-based partnership organisations with an economic and community development remit were developed from 1991 onwards, as part of a network of over a hundred urban and rural local partnerships implementing a variety of EU and national programmes (Walsh, 2001). Unlike all four jurisdictions within the UK, the Republic of Ireland does not have a national urban regeneration strategy within which local programmes may be contextualised. National strategies and policies that influence urban regeneration include the various Urban Renewal Scheme; the *National Spatial Strategy*; the *National Anti-Poverty Strategy*; the social partnership agreements and the *National Development Plan*.

Sustainable Development and Sustainable Participation

The cultural and political changes of the 1960s which brought about the beginnings of public participation in urban regeneration also led to greater awareness of ecological issues including the environmental costs of development and subsequently the concept of sustainable development (Ekins, 1992). The Bruntland Report defines sustainable development as "development that meets the needs of the present without compromising the ability of future generations to meet their own needs" (World Commission on Environment and Development [WCED], 1987, p. 8). The report explains that "sustainable development is not a fixed state of harmony, but rather a process of change" (WCED, 1987, p. 9) which includes economic, social and environmental factors; and that implementation requires "a political system that secures effective citizen participation in decision-making" (WCED, 1987, p. 65).

Commitment to public participation was reinforced at the United Nations Conference on Environment and Development in Rio (United Nations, 1992a), including the Agenda 21 action plan for sustainable development at global, national and local levels (United Nations, 1992b). Sustainable development strategies were subsequently produced by both the United Kingdom (HM Government, 1999) and the Republic of Ireland (DoE, 1997); again, the importance of public participation was emphasised. Under devolved government, Northern Ireland is developing its own sustainable development strategy as part of the revision of the UK's strategy (DEFRA, 2004).

Therefore public participation is considered an essential part of sustainable development strategies. The question of what constitutes 'sustainable' participation is, however, left uncertain. The interaction between state agencies and the organisations of civil society is mediated through participation structures. Following the Bruntland Report, sustainable participation could be defined as: participation processes and outcomes that address the interests and concerns of present participants without damaging future representation. However, this definition omits the importance of the dynamic aspect of effective participation and the complexity of factors and interests involved. In order to be able to cope with the 'process of change' inherent in sustainable development (WCED, 1987, p. 9), participation structures must be capable of facilitating negotiations between state agencies and the organisations of civil society. The structure and membership of partnerships may themselves change over time as part of a response to changing circumstances, in order to retain the ability to negotiate change. Therefore a better definition of sustainable participation might be: participation processes capable of negotiating optimal outcomes to currently changing circumstances, without damaging the ability to respond to future change.

The State, Civil Society and Sustainable Participation

It is well recognised that state organisations need to obtain legitimation of their actions by civil society, in order to maintain social and economic stability. In urban regeneration programmes, legitimation is sought through public participation in partnership processes. Legitimation of state activity assists with the maintenance of hegemony, defined here as: the importance of a consensual social structure which allows power to be maintained through the establishment and upholding of dominant ideological positions within the institutions of civil society (Gramsci, 1971; Simon, 1991). Alternatively, counter-hegemonic positions may be developed within civil society, to challenge the state agenda.

Jessop (1997) describes the implementation of 'hegemonic projects' which help to stabilise society and therefore create or maintain conditions for capital accumulation, through achieving "relative unity of diverse social forces" and resolving "conflicts between particular interests and the general interest" (Jessop, 1997, p. 62). Public participation in urban regeneration, within both Irish jurisdictions, is a hegemonic project. The nature of participation may be contested, but its practice is not. The establishment of regeneration partnerships has secured 'relative unity' between the organisations of the state and civil society, has contributed to a strong ideology of common interests, and has provided an arena in which social conflict can be managed.

Factors Affecting Sustainable Participation: A Theoretical Framework

The following theoretical framework attempts to bring together the elements that affect public participation in urban regeneration. Participation is theorised as a dynamic site of complex and uneven interaction between the state and civil society. The framework acknowledges that factors affecting this interaction may be found at the micro-, meso- and macro-levels of society and may be social or economic in origin. No one theory is regarded as adequate for this purpose, although the micro- and meso- theories represent a more detailed consideration of aspects of the mode of regulation within the macro- level framework of regulation theory. The three components of such a framework put forward here are:

- Micro-level power relationships, primarily within formal structures, as explored by Lukes (1974) and Clegg (1989);
- Networks and regimes, at the micro- and meso-levels, primarily outside formal structures (Rhodes, 1997; Stoker, 1995; Dowding, 2001);
- Macro-level forces which influence and maintain economic, social, political and cultural stability, through the structure of regulation theory (Jessop, 1990; Painter & Goodwin, 1995; Jessop, 1997).

Of the many writers on power relationships (surveyed by Clegg, 1989), two have been selected for this analysis. Lukes (1974) created a three-stage model of the use of power: the first dimension, focussing on observed behaviour and decision-making; the second, with added attention to non-decisions, hidden conflict and the 'mobilisation of bias' by control of the agenda; and the third, including latent conflict due to a dominant value system which removes some potential issues from consideration. Clegg (1989) added to the debate by producing a typology of circuits of power, where power is located in either social relations (where Lukes' analysis is concentrated), rules, or systems of physical domination; power moves from one circuit to another depending upon the empirical situation, often encountering resistance.

The concepts of networks and regimes provide a second perspective on participation, this time focused upon interactions between the state and civil society that take place outside formal structures, primarily at micro- and meso-levels. Networks are relationships between interest groups and government involving resource exchange (Rhodes, 1997). At the micro- level, "networks arise from and are sustained by the relationships between individuals over some shared concern, belief or value" (Skelcher et al., 1996, p. 8). Networks have increased in number and importance as part of the move from government to governance through the widening of the number and types of agencies involved in the administration of state-sponsored initiatives, in Britain and Ireland (Adshead & Quinn, 1998; Rhodes, 1997).

Regimes and networks are related concepts. Regimes consist of "coalitions of the elected, business and other pressure organisations, and may include important bureaucratic and professional groups" (Dowding, 2001, p. 8). The role of the state within regimes is sometimes to direct, but more usually to mediate and to participate in negotiation and bargaining with other agencies (Stoker, 1995). The differences between networks and regimes are that regimes usually, but not always, encompass a wider range of interests; last for longer and survive personnel changes; and have a clear policy direction.

Lastly, the framework acknowledges the impact of wider forces upon area-level programmes through the application of a regulation theory framework to review economic and social forces including hegemonic aspects. The basis of the regulation approach is that economic change is connected to and influenced by change in political, cultural and social factors (Painter, 1995). The key analytical categories are the regime of accumulation and the mode of regulation. The regime of accumulation is the organisation of production, management and consumption to create surplus value. The mode of regulation includes the social relationships and laws that create the circumstances in which capital accumulation can continue, for example legislation, politics, industrial relations, and social expectations (Amin, 1994). Most variants of regulation theory also include a third element that deals in some way with hegemonic concepts that help to stabilise the regime of accumulation and the mode of regulation (Jessop, 1990). Regulation theory is most productively approached as a process rather than as a static model (Painter & Goodwin, 1995).

Introduction to the Case Studies

The seven-year North Belfast Housing Strategy was launched in 2000 and covers a patchwork of Catholic and Protestant communities in the most deprived parts of North Belfast. The implementing agency is the Northern Ireland Housing Executive (NIHE), the region's strategic housing agency and Northern Ireland's largest social housing landlord. The Strategy aims to increase housing supply; make better use of existing stock; improve the NIHE's housing stock; sustain and improve private housing areas; and promote regeneration and social inclusion through partnerships (NIHE, 2000). The Strategy's programme was based upon an analysis of housing need which showed that the two main communities in North Belfast had different requirements, with Catholics needing more social housing due to a growing number of families; and Protestants needing improvements to existing stock including assistance in the private sector (NIHE, 2000). The implementation of the Strategy has taken place against a history and background of sectarian violence (Jarman, 2002).

The second area, Ballymun, is located five miles north of the centre of Dublin and contains 5,000 dwellings, of which just over half are system-built flats. The estate was built in the mid-1960s in response to a severe housing crisis, and suffered management and social problems from its early days (Power, 1997; Somerville-Woodward, 2002). Ballymun Regeneration Ltd (BRL), a company wholly owned by Dublin City Council (DCC), was established in 1997. The regeneration Masterplan (BRL, 1998) aims to construct low-rise flats and houses to replace the original flats; improve employment and training opportunities; provide better schools, a town centre and new shopping centre; and improve neighbourhood identities reinforced by design of new facilities. The programme will continue until at least 2012. Three organisations have key responsibilities within the programme: Ballymun Regeneration Ltd (BRL) as the development agency, Dublin City Council (DCC) as the main social housing landlord; and the Ballymun Housing Task Force (BHTF) as the community liaison organisation and facilitator of the consultation process. The context of the planning and implementation of the Ballymun programme is the Republic's 'Celtic Tiger' economic boom of the 1990s and its consequences, which are increasingly acknowledged to have included an increase in inequality along with material benefits for much of the population (Kirby, 2002).

The Consultation Processes

The results of the case study research in North Belfast and Ballymun provided rich empirical data. In North Belfast, consultation and implementation were almost completely separate processes within the Housing Strategy. Two new consultation bodies were established: a community forum for the whole of the Strategy's area, to consult with representatives from both communities; and an inter-agency group, to share information with statutory agencies and other city-wide interests. Progress on implementation was reported to the consultation bodies, but detailed monitoring took place elsewhere. Other community activity in North Belfast was fragmented both strategically and operationally. The area's many community groups mainly served single-community areas although many were also engaged in separately funded cross-community work. Networks within community boundaries were strong, and there were some important contacts between the two communities. Yet, funding for community groups was short-term and uncoordinated. Interviews with participants in the Strategy from all sectors revealed generally negative views of partnership at work. Partnership forums were seen by community workers as useful for obtaining information rather than for genuine joint working. However, opinions of the Housing Strategy were more positive than those of partnership working in general.

In Ballymun, the regeneration programme included a complex consultation and decision-making structure resulting at least partly from the fact that key responsibilities were divided between three organisations. There was a high degree of overlapping membership between the decision-making bodies of the three organisations, which assisted with information sharing and debate. However, as in North Belfast, implementation structures remained separate. The consultation process consisted of five Estate Forums covering neighbourhoods within the estate, reporting to the Task Force Board and subsequently to the BRL Board. Unlike North Belfast, consultation sometimes delayed the implementation of the programme; not through involvement in the detail, but through prolonged and often acrimonious debate at the planning stage. The large number of agencies and community groups working in Ballymun created a complex and fragmented network of organisations and individuals. It was estimated that there were between 100 and 200 Ballymun community groups at the time of the fieldwork, but there was no umbrella organisation that could represent all groups within the regeneration process.

Most Ballymun interviewees had a very clear vision of how partnerships should work, for example: openness, transparency and honesty in decision-making; working through consensus; and working together as equals. However, everyone acknowledged that the practical experience of partnerships fell far short of their ideals and included poor communication; much anger; and tiring, stressful meetings. Opinions of the regeneration programme itself were cautiously positive, with the exception of one Estate Forum who opposed the programme as they felt it was not in their interests.

Other Factors Affecting Participation

The implementation of the North Belfast Housing Strategy took place against a background of entrenched conflict dating back to the early nineteenth century (Bardon, 1992) that had resulted in a divided civil society and acute territorial divisions; North Belfast residents accounted for 16 per cent of all deaths during the 'Troubles' (Fay et al., 1998). Sectarian tensions during the fieldwork period included regular interface violence, several sectarian murders, and the Holy Cross school protests. A feud between the two main loyalist paramilitary organisations created additional divisions in Protestant areas. Acknowledgement of the depth of local sectarian conflict occurred several times in interviews but never in public forums.

The division of civil society in North Belfast was acute, with the two main communities occupying separate spatial, political and to lesser extent economic spheres as discussed by Murtagh in the previous chapter (Shirlow, 2001). The activities of community groups reflected this; so did the political system, with most political parties seeking their votes from one community only. The assessment of housing need followed the dual nature of civil society by providing gains for each side. The divided civil society was exemplified by territorial divisions, which were linked to community identity and caused decisions about land use to be intensely political (Murtagh, 2002), especially as the high level of housing need in Catholic communities was challenging territorial boundaries. The NIHE's position on disputed territory was that 'adjustment' would only be possible with the consent of both communities (NIHE, 2000).

Another aspect of North Belfast's divided civil society that impacted on the Strategy's consultation process was that the two communities were not equally represented. Strategy participants confirmed the widely held perception of weak community development in Protestant areas (CDPA, 1991), Protestant 'alienation' (Dunn & Morgan, 1994) and 'defeatism' (Finlay, 2001) in response to social and economic changes that had removed their privileged position in the labour market, and as a response to some elements of the Northern Ireland peace process. The Strategy's community forum was an unusual example of a cross-community consultation body in North Belfast, but attendance from Protestant groups was very much lower than that of Catholics. Disappointment was expressed about this, but no examination was made of structural issues that might be causing it, such as different patterns of community organisation within Protestant communities (Langhammer, 2003).

Turning to Ballymun, the two other themes that arose from the research were the importance of tackling social exclusion, and the dynamic of distrust within the consultation process. Ballymun was a severely socially excluded community and one of the most deprived areas in the country. The regeneration programme aimed to re-integrate Ballymun into the surrounding suburbs. The programme of new house-building was an important part of reducing the social exclusion of the area and the new homes were in general appreciated by residents, some of whom, when interviewed, indicated their willingness to become involved in community activities for the first time. There were, however, two exceptions to the approach of re-integration. Firstly, there was no strategy or ownership by a statutory agency of the problem of anti-social behaviour on the estate, which looked likely to continue to set Ballymun apart from its neighbours. Second was the emphasis on community development in Ballymun. Unlike the residents of the surrounding districts, those living in Ballymun were considered to need support in sustaining community activity and involvement.

The implementation of Ballymun's regeneration programme was marked by a dynamic of distrust within the consultation process, which differed in degree depending upon the area or projects involved. The level of distrust was puzzling until it was placed in a historical context. The period from the early 1980s until the mid-1990s saw the gradual development of trust between Dublin Corporation and Ballymun's community sector (Somerville-Woodward, 2002). The working relationship was undermined at the start of the regeneration programme by the creation of a new agency, BRL. In addition, there was lasting resentment that BRL had commissioned outside consultants to carry out the Masterplan consultation in 1997 rather than working through local community groups. A second reason for distrust was the opposition of one Estate Forum to the programme on the grounds that it would not improve their quality of life. This area did not include any of the large flat blocks that were being demolished and the Forum resented what they saw as use of their green spaces to provide housing for people from other parts of Ballymun. The impact of these factors on the regeneration programme was that the consultation process was often slow and conflict-ridden, especially within the Forum that opposed the regeneration programme.

Public Participation in Urban Regeneration in Ballymun and Belfast

Patterns of public participation in Irish urban regeneration programmes indicate that consultation structures are not concerned primarily with programme implementation. Rather, the reasons for the establishment of partnership structures are to promote community development and to encourage community legitimation of the programme. In North Belfast, community development was encouraged through the practice of 'community relations' as part of the management of conflict. The Strategy's consultation structure sought to improve community relations through requiring community representatives to work together at a time when the two communities were still in severe and violent dispute. Conflict was managed through a lack of discussion about contentious issues and by failing to consider alternative structures in order to encourage greater participation by Protestant representatives. The Strategy itself was legitimated by both communities because it worked with the duality of civil society and offered something for each community.

In Ballymun, community development took the form of the encouragement of organised community activity as part of the management of social inclusion. The Masterplan stated that the programme aimed to change Ballymun's social and economic structure "to reflect more closely a well balanced community which is an integral part of North Dublin" (BRL, 1998, p. 8). However, the paradox that the surrounding areas of North Dublin did not contain such a high level of community development activity was not remarked upon. Legitimation of Ballymun's programme was less whole-hearted than would have been expected, given the Republic's national level ideology of social partnership. The objectives of the programme were supported by four out of five of the Estate Forums, but individual projects were sometimes opposed in all areas.

Participation and Hegemony

It was proposed earlier in the chapter that public participation in urban regeneration is a hegemonic project in both Irish jurisdictions because: the effectiveness of participation is not questioned (even when the practice leaves participants dissatisfied); participation structures have secured 'relative unity' between the state and civil society through contributing to an ideology of common interests; and because participation, through partnership structures, has provided an arena for the management of social conflict.

In both case study areas, there was an assumption by all participants that public participation was essential for the success of the programmes. In neither area was there a strong counter-hegemonic discourse. Equally, in both areas the ideology of common interests centred on the programme implementation documents. In North Belfast, there was widespread support for the Strategy (NIHE, 2000) and its five objectives were used as a framework to report progress to consultation forums. In Ballymun, the status of the Masterplan (BRL, 1998) was more ambiguous. As previously indicated, one Estate Forum rejected the regeneration programme and hence the contents of the Masterplan. Others sought to question its contents and were in some cases told that details had changed in the course of the redevelopment. The 'relative unity' between the state and civil society described by Jessop (1997) was less in evidence in Ballymun. Finally, partnership structures in both case study areas were used to manage social conflict, with limited success. In North Belfast many issues were not discussed openly in any forum and in Ballymun serious disagreements were often discussed and then referred on to the planning process. Therefore aspects of a hegemonic project were more completely in evidence in North Belfast than in Ballymun, but in each case some hegemonic elements were present.

Understanding Participation

This section considers issues of macro-level forces, networks and regimes, and power, which affected participation processes in the case study areas, as laid out in the theoretical framework earlier in the chapter.

The North Belfast Housing Strategy

As indicated in the Introductory chapter, Northern Ireland's regime of accumulation has been in stable transition from a manufacturing-based to a service-based economy in recent years. Notwithstanding this stability, aspects of the mode of regulation remained volatile, especially the duality of civil society and the technocratic approach to the delivery of public services. Macro-level economic change from manufacturing, with mainly Protestant workforces, to the more balanced distribution and employment patterns of the contemporary service industries, is regarded as a key factor in what has been described as Protestant 'alienation' (Dunn & Morgan, 1994). The consequences of the duality of civil society severely affected the Strategy's consultation process and also affected the presentation of the needs analysis on which it was based. The technocratic approach to public administration caused the Strategy's consultation and implementation process to remain separate and for programme targets to be unaffected by the consultation process.

The influence of networks and regimes was also linked to the technocratic approach within the mode of regulation. The key regime was that of regional governance, incorporating the UK government (politicians and civil servants); regional government departments, politicians; and public sector agencies including the NIHE. There was concern within the regime at the increasing influence of local politicians from the Northern Ireland Assembly, as North Belfast's representatives were perceived by bureaucrats as being concerned only with the benefits to their own community. Networks were also affected by the duality in civil society, which was at the heart of the mode of regulation. The key networks included: a strong housing professional network, divided into two parts, one based around the NIHE and one around housing associations (including developers and building contractors); and three community-based networks, one for each community and one based on cross-community contacts. There was also a degree of overlapping membership that facilitated informal contacts outside the Strategy's consultation process.

At the micro- level, power relationships were based upon the mobilisation of bias in favour of the Strategy as a response to the dual analysis of housing need, and a dominant value system that endorsed a cross-community approach to consultation. There were no meaningful discussions about community conflict or about facilitating greater inclusion of the Protestant community. In many cases conflict appeared to be expressed by failing to attend rather than by bringing conflict into the group. The shadow of power exercised through physical domination was part of the context of the Strategy's implementation, among which were the activities of paramilitary organisations in North Belfast affecting the location of new social housing by maintaining territorial boundaries, as well as causing civil unrest, homelessness and property blight in interface areas.

The Ballymun Regeneration Programme

In contrast in the Republic of Ireland, the regime of accumulation has undergone a rapid transformation from a weak manufacturing base to a globalised branch plant service-based economy, along with a significant retained agricultural sector. The mode of regulation was stable during the study period was stable, underpinned by the national social partnership structure which included a commitment to tackling social exclusion, along with a political emphasis on the importance of low corporate and personal taxation. A still buoyant economy, skills shortages and low unemployment had led to an incentive to re-integrate the reserve army of labour in places such as Ballymun and had therefore been helpful for the context of the regeneration programme.

Networks and regimes in this area were complex and significant. The most influential regime was at city-wide level and included: national government officials and politicians; DCC (including politicians) and BRL; the BHTF and other local voluntary organisations. Local networks were complex and overlapping, including: a statutory and semi-state network including DCC staff, BRL, and other statutory agencies; a community network including DCC staff, the Estate Forums, the BHTF, and other local community groups; and a voluntary sector network based on the national level social partnership structure. The way in which local networks had formed and changed over the years had been an important aspect of the underlying distrust between civil society and the state in Ballymun, with BRL still a relative newcomer to the complex environment.

Micro- level power relationships in Ballymun had also been shaped by local history and included a fair amount of open conflict. The mobilisation of bias in the consultation process was towards the benefits of the programme. The dominant value system strongly emphasised the importance of consultation and community development; however, many groups felt excluded from the consultation process. There was resistance to the dominant values from the Estate Forum, which opposed the programme, including the use of the consultation process and the planning appeals process to cause delays. Power through physical protest had also been expressed at times through non-violent direct action against building contractors.

Towards Sustainable Participation?

This chapter concludes by considering how the reported research findings and analysis may contribute towards a debate about sustainable participation processes in Irish urban regeneration programmes, with sustainable participation defined earlier as: participation processes capable of negotiating optimal outcomes to currently changing circumstances, without damaging the ability to respond to future change.

Each case study consultation structure was set up to monitor and receive information on the progress of a written programme being implemented to a set of targets, hence scope for negotiation was constrained. In North Belfast, the programme was to be implemented in a balanced way, to ensure that each community could be seen to have gained, and any changes to the programme had to take that overview into account. The consultation structures were not open to change in order to maximise participation, although good informal links and a series of local meetings meant that the NIHE was able to obtain the views of both communities despite problems within the community forum. The failure to discuss many important local issues kept the forum in operation but compromised its effectiveness and contributed to the disillusion of participants, thus arguably damaging the ability to respond to future change. To summarise, the consultation process in North Belfast was not flexible and open enough to negotiate change, primarily due to the wider socio-environmental context within which it operated.

In Ballymun, more negotiation about the details of individual schemes took place, especially as part of consultation on planning applications. The consultation process was reviewed during the case study research period, with the result that the new proposed structure would increase collective responsibility for decisions and clarify rules. The expression of open conflict, and the attempts to resolve at least some elements of disagreement, was paradoxically a positive aspect when compared to the avoidance of issues in North Belfast, although it contributed to a sense of frustration with the process. The consultation process in Ballymun has moved towards an ability to negotiate change, by acknowledging that structures might need to change and that conflict is sometimes inevitable. Therefore, although neither case study consultation process could be described as 'sustainable', Ballymun was the more promising of the two.

The results of the research raise several points about the elements required for sustainable public participation. It is helpful to see participation as a site of complex interaction between organisations of the state and of civil society, within which ideology will influence behaviour and choices. External factors will have an impact on the process, for example economic change and patterns of governance, and there may be times when programmes or implementation processes need to be changed to reflect this. Greater awareness and articulation of the historical factors that have influenced relationships between participants would be helpful. There should also be a greater acknowledgement of the effects of other agendas, such as political or administrative priorities that lie outside the programme, and a shared understanding of the way in which bureaucratic regimes and networks often exclude community representatives. Community planning exercises should include an analysis of the ways in which power works within consultation structures.

The essence of sustainable participation is the ability to negotiate change, now and in the future. The intention of state agencies to build community 'capacity' within urban regeneration programmes should be undertaken in the context of a deeper understanding of the inequalities between the state and civil society and the ways in which consultation processes operate.

References

Adshead, M. & Quinn, B. (1998) 'The Move from Government to Governance: Irish development policy's paradigm shift', *Policy and Politics,* Vol. 26(2), pp. 209-225.

Amin, A. (1994) 'Post-Fordism: Models, Fantasies and Phantoms of Transition' in Amin, A. (ed.) *Post-Fordism: A Reader*, Oxford: Blackwell.

Bardon, J. (1992) *A History of Ulster*, Dundonald: Blackstaff Press.

BRL (1998) *Masterplan for the new Ballymun*, Dublin: Ballymun Regeneration Ltd.

CDPA (1991) *Community Development in Protestant Areas*, Belfast: Community Development in Protestant Areas Steering Group.

Clegg, S. (1989) *Frameworks of Power*, London: Sage.

Curtin, C. & Varley, T. (1995) 'Community Action and the State', in Clancy, P., Drudy, S., Lynch, K. & O'Dowd, L. (eds) *Irish Society: Sociological Perspectives*, Dublin: Institute of Public Administration.

DEFRA (2004) *Taking It On: Developing UK sustainable development strategy together; a consultation paper*, London: Department for the Environment, Food and Rural Affairs.

DoE (1997) *Sustainable Development: a strategy for Ireland*, Dublin: Department of the Environment.

Dowding, K. (2001) 'Explaining Urban Regimes', *International Journal of Urban and Regional Research*, Vol. 25(1), pp. 7-19.

DSD (2003) *People and Place: A Strategy for Neighbourhood Renewal*, Belfast: Department for Social Development.

Dunn, S. & Morgan, V. (1994) *Protestant Alienation in Northern Ireland: A Preliminary Survey*, Coleraine: University of Ulster.

Ekins, P. (1992) *A New World Order: Grassroots movements for global change*, London: Routledge.

Fay, M.T., Morrissey, M. & Smyth, M. (1998) *Mapping Troubles-related Deaths in Northern Ireland 1969-1998*, Derry: INCORE.

Finlay, A. (2001) 'Defeatism and Northern Protestant "Identity"', *Global Review of Ethnopolitics*, Vol. 1(2), pp. 3-20.

Gramsci, A. (1971) *Selections from the Prison Notebooks*, (ed. Q. Hoare, Q. & G. Nowell Smith), London: Lawrence and Wishart.

Greer, J. (2001) *Partnership governance in Northern Ireland: improving performance*, Aldershot: Ashgate.

HM Government (1999) *A Better Quality of Life: a strategy for sustainable development for the UK*, Cmd 4345, London: Stationery Office.

Hughes, J., Knox, C., Murray, M. & Greer, J. (1998) *Partnership Governance in Northern Ireland: the Path to Peace*, Dublin: Oak Tree Press.

Imrie, R. & Raco, M. (2003) 'Community and the changing nature of urban policy' in Imrie, R. & Raco, M. (eds) *Urban Renaissance?: New Labour, community and urban policy*, Bristol: The Policy Press.

Jarman, N. (2002) *Managing Disorder: Responding to Interface Violence in North Belfast*, Belfast: Office of the First Minister & Deputy First Minister (Equality Directorate Research Branch).

Jessop, B. (1990) 'Regulation theories in retrospect and prospect', International *Journal of Urban and Regional Research*, Vol.19(2), pp. 153-216.

Jessop, B. (1997) 'A Neo-Gramscian Approach to the Regulation of Urban Regimes: Accumulation Strategies, Hegemonic Projects, and Governance', in Lauria, M. (ed.) *Reconstructing Urban Regime Theory: Regulating Urban Politics in a Global Economy*, Thousand Oaks: Sage.

Kirby. P. (2002) *The Celtic Tiger in Distress: Growth with Inequality in Ireland*, Basingstoke: Palgrave.

Langhammer, M. (2003) *Cutting with the Grain: Policy and the Protestant Community – what is to be done?*, Unpublished paper prepared for the Northern Ireland Office.

Lukes, S. (1974) *Power: A Radical View*, London: Macmillan.

McCready, S. (2001) *Empowering People: Community Development and Conflict 1969 – 1999*, Belfast: The Stationery Office.

Muir, J. (2004) *The state and civil society in urban regeneration: the representation of local interests in area-based urban regeneration programmes*, Unpublished PhD thesis, School of the Built Environment, Faculty of Engineering, University of Ulster.

Murtagh, B. (2001) 'The URBAN Community Initiative in Northern Ireland', *Policy and Politics*, Vol. 29(4), pp. 431-446.

Murtagh, B. (2002) *The Politics of Territory: Policy and Segregation in Northern Ireland*, Basingstoke: Palgrave.

NIHE (2000) *The North Belfast Housing Strategy*, Belfast: Northern Ireland Housing Executive.

Painter, J. (1995) 'Regulation Theory, Post-Fordism and Urban Politics', in Judge, D., Stoker, G. & Wolman, H. (ed.) *Theories of Urban Politics*, London: Sage.

Painter, J. & Goodwin, M. (1995) 'Local governance and concrete research: investigating the uneven development of regulation', *Economy and Society*, Vol. 24(3), pp. 334-356.

Pateman, C. (1970) *Participation and Democratic Theory*, Cambridge: Cambridge University Press.

Power, A. (1997) *Estates on the Edge: The Social Consequences of Mass Housing in Northern Europe*, Basingstoke: Macmillan.

Rhodes, R.A.W. (1997) *Understanding Governance: Policy Networks, Governance, Reflexivity and Accountability*, Buckingham: Open University Press.

Robson, T. (2000) *The State and Community Action*, London: Pluto Press.

Shirlow, P. (2001) 'Fear and Ethnic Division', *Peace Review*, Vol. 13(1), pp. 67-74.

Simon, R. (1991) *Gramsci's Political Thought: An Introduction*, London: Lawrence & Wishart.

Skelcher, C., McCabe, A. & Lowndes, V. (1996) *Community networks in urban regeneration: 'It all depends who you know...!'*, Bristol: Policy Press.

Social Exclusion Unit (2001) *A New Commitment to Neighbourhood Renewal: National Strategy Action Plan*, London: Cabinet Office.

Somerville-Woodward, R. (2002) *Ballymun: A History, Volume 2, c.1960-2001*, Dublin: Ballymun Regeneration Ltd.

Stoker, G. (1995) 'Regime Theory and Urban Politics', in Judge, D., Stoker, G. & Wolman, H. (ed.) *Theories of Urban Politics*, London: Sage.

United Nations (1992a) *Report of the United Nations Conference on Environment and Development, Annex I: Rio Declaration on Environment and Development*, http://www.un.org/documents.

United Nations (1992b) *Agenda 21*, http://www.un.org/esa/sustdev/documents.

Walsh, J. (2001) 'Catalysts for change: public policy reform through local partnership in Ireland', in Geddes, M. & Benington, J. (eds) *Local Partnerships and Social Exclusion in the European Union: new forms of local social governance?*, London: Routledge.

Walsh, J., Craig, S. & McCafferty, D. (1998) *Local Partnerships for Social Inclusion?*, Dublin: Oak Tree Press.

World Commission on Environment and Development (1987) *Our Common Future*, Oxford: Oxford University Press.

Chapter 12

Active Citizenship: Resident Associations, Social Capital and Collective Action

Paula Russell, Mark Scott and Declan Redmond

Introduction

This chapter explores the nature of residents' associations and their role in different types of communities in the city of Dublin. In Irish society, as elsewhere, there is a growing interest in the role of civil society and the potential to create deeper and more embedded democratic culture and practice. In Ireland the development and strengthening of the organisations of civil society, particularly voluntary and community groups, is considered to be of importance and has been given credence by the publication of a government white paper on Voluntary and Community Activity by the Irish Government (Government of Ireland, 2000). More recently, the Democracy Commission has highlighted the centralised decision making system which prevails in Ireland as undermining the legitimacy of local democracy and militating against local participation in civic society bodies such as residents' associations (Democracy Commission, 2004, p. 6). It is these residents' associations that are the focus of the research on which this chapter draws. The hypothesis is that residents' associations are one of the few ways that private neighbourhoods organize at a local level and interact collectively with the local state and that they provide a forum for community development and neighbourhood identification.

The chapter explores the nature of the relationship between residents' associations and the local state and in particular the manner in which residents' associations act as the bastion of opposition to what is perceived as unwelcome development in their community. The first part of the chapter reviews some of the literature relating to social capital as a means of exploring the potential of residents' associations to act as network builders in their communities, bonding the community and providing a sense of shared endeavour, while also linking the community to other actors and institutions. The second part of the chapter outlines some of the empirical research which we have carried out which explores the nature of residents' associations and their role in building bonding and bridging social capital at neighbourhood level.

What is Social Capital?

The concept of social capital has a long history and can be traced to the work of the classical sociologists including Durkheim, Marx and others (Portes, 1998; OECD, 2001). More recently, it has been developed by sociologists such as Bourdieu (1986), Coleman (1988) and Portes (1998). However, it is Putnam (1993, 2000) who has popularised the term and has sparked the most recent theoretical and popular debate around the concept and it is to his definition of the term that we will later return. The term social capital was initially developed to temper the analysis of capital in purely economic terms. Sociologists and others argued that classical analyses of capital as purely financial and physical ignored the value that lies both in individual's knowledge and skills (cultural capital) and in the social networks and shared values that facilitate cooperation between actors (social capital). While there are differences in theorists understanding of social capital, there is one central feature in all of their definitions, that is that social capital relates to the relationships between individuals. For Putnam "social capital refers to connections among individuals – social networks and the norms of reciprocity and trustworthiness that arise from them" (Putnam, 2000, p. 19).

In unpicking this definition of social capital, it is clear that it refers to the manner in which people interact through social networks and other social relationships. This interaction is facilitated by the trust that people place in each other, by norms of reciprocity and a sense of mutual obligation. It is an expectation that if I do a favour for someone that at some undefined stage in the future this favour will be reciprocated. This is the type of understanding expressed in the old adage that 'one good turn deserves another'. The appeal of social capital and the reason why it is of key interest to academics and policy- makers alike is the belief that higher levels of social capital can lead to positive outcomes in a range of areas. These positive outcomes include better governance, reduced crime, healthier and better-educated communities and enhanced economic development (Office for National Statistics, 2001; Field, 2003). As Putnam states: "an impressive and growing body of research suggests that civic connections help make us healthy, wealthy and wise" (Putnam, 2000, p. 287).

For Putnam, social capital is a collective good that can be possessed by neighbourhoods, cities or even nations and it is exemplified by civic engagement. At neighbourhood or community level there is a sense that the interaction and trust that exemplifies social capital and which knits or glues communities together will facilitate good neighbourliness and will inspire collective action to achieve shared goals. Thus, this concept of social capital may help us understand the nature of role of residents' associations, why they develop and how they operate?

Crucial to understanding social capital in more depth are the different types or faces (Briggs, 1998, 2004) of social capital that have been identified. Putnam distinguishes between two types of social capital – bonding and bridging – which are central to the understanding of the term, while Woolcock (2000) includes a third dimension of linking social capital. Bonding social capital relates to ties within communities and organizations which bring together those who are similar to each other on the basis of ethnicity, education, interests, social background or any other dimension (Healy, 2004). It creates strong in group identity and loyalty and is useful for supporting specific reciprocity and mobilizing solidarity. By its nature it tends to be exclusive. This is the social capital that Briggs (1998) describes as a social support that helps people 'get by' or cope.

Bridging Social Capital on the other hand, refers to the ties which link more diverse people, linking those who are not similar in social background, ethnicity etc. Bridging social capital is therefore more outward looking and cross cutting, it encompasses the weaker ties which Granovetter (1973, 1995) has pointed out are important in job searches in areas outside those where family and friends were already employed. Briggs describes this as "social leverage" or social capital that helps a person get ahead (Briggs, 1998, p. 178). Bridging social capital, as a result, can be understood as relating to the manner in which communities can link to the external environment (Taylor, 2003). For many authors bridging social capital is seen as a potentially more powerful form of capital (Larsen et al., 2004; De Filippis, 2001).

Linking social capital is similar to bonding social capital, but more specifically relates to relations between individuals and groups at different levels of social status or power (OECD, 2001, p. 42). Woolcock (1998) sees linking social capital as a means for the community to leverage resources, ideas and information from formal institutions outside the community. It is the collaborative or external social capital that Purdue (2001) relates to the links that community leaders make with external agencies such as banks, local authorities and funding bodies and which may be of importance to residents' associations. These three different types or dimensions of social capital allow for a more coherent understanding of the possible range of outcomes of differing combinations of these types (Woolcock and Narayan, 2000).

Another important facet in understanding social capital is to understand how it is related to other forms of capital. While some authors make a concerted attempt to distinguish social capital from other forms of capital such as cultural and economic capital (Mohan and Mohan, 2002), it is important to note that while these forms of capital are not the same and are not necessarily correlated, they may be instrumental in understanding how different communities have differential access to social capital. Thus, the complexion of the social capital that exists in different communities, particularly bridging social capital, may be very different depending on the other forms of capital which are possessed by the individuals in that community. Bourdieu's (1986) understanding of social capital is useful in this regard.

In contrast to Putnam, Bourdieu (1986) does not see social capital as a product of collective action, rather he sees it as an individual resource. While he believes that social capital inheres in people's networks and relationships, it is realised by individuals. Bourdieu's writings on social capital were related to his work on social hierarchy and the important role which cultural capital played as an asset which groups used to maintain superiority over others. Therefore, he understood social capital as one of the means by which people maintained their position, by utilising their networks and connections. For Bourdieu, social and cultural capital are rooted in economic capital and ultimately can be reconverted to economic capital. The transformation of social capital to economic capital, however, is not straightforward. Bourdieu pointed out that the time horizons in which an individual can realise the benefits or social capital are long, (given that relationships have to be established and maintained for a long time, and may or may not be of use to us). Likewise, the reciprocity you may expect can never be guaranteed. Bourdieu thus believed that the expenditure of time and energy needed to acquire social capital tends to be underpinned by economic capital.

Bourdieu's conceptualisation of social capital thus highlights the importance of ensuring that social capital is not separated from underlying economic capital and that the power that inheres in economic and cultural capital are recognised. Throughout his work Bourdieu emphasises the role played by different forms of capital in the reproduction of unequal power relations, and the manner in which the dominant class can ensure that the capital which they possess is that which is most valued by others in society. His conception of social capital is thus useful in understanding the manner in which the residents' associations in affluent areas are in a position to utilise their own economic and cultural capital to underpin social capital, both for their own benefit and the benefit of the area as a whole. In this way we can understand why Hall (1999) writing about the British situation, can observe that social capital is not distributed evenly among the population: it is disproportionately, a middle class phenomenon and the preserve of those in middle age.

Although Bourdieu clearly sees social capital as realised by individuals, the manner in which social capital is underpinned by cultural and economic capital has been shown by others to be an important factor in promoting collective action. Research by a range of authors documented by Larsen et al. (2004) shows that higher levels of social status are a key predictor of collective action. Larsen et al.'s (2004) own research shows that people in affluent neighbourhoods, and with higher social status (measured as educational attainment) are more likely to act on neighbourhood environmental problems than those in less affluent neighbourhoods and with lower levels of educational attainment. Empirical research of neighbourhood action in middle class neighbourhoods in London has also illustrated the linkages that exist between social capital and cultural capital, such as education and knowledge. Butler and Robson's (2001) case study of Telegraph Hill in London shows the manner in which the networks of social capital which existed through the residents' association achieved the general upgrading of the area by acting collectively on their cultural capital. The elements of cultural capital which were important in relation to environmental change were documented as including a range of social skills: "detailed knowledge, case preparation, articulacy and social confidence – all advantages in dealings with local government officers and other key institutional personnel"

(Butler and Robson, 2001, p. 2159). The middle class residents of Butler and Robson's study utilised social capital to ensure cultural capital for their children, to protect the value of their property investment and to protect their efforts to gentrify the three areas of London studied.

In studies of participation it has also been found that public forums are often dominated by established middle class people, those who possess considerable cultural and economic capital. Therefore, as Vigar and Healey (2002, p. 522) comment:

> … opportunities to participate in governance processes are typically taken by those with power or who already have formal or informal access to decision-making nexuses. Well-resourced groups tend to be favoured by such processes. Thus, for example [in relation to managing environmental conflict], forms of environmental racism or classism can emerge …

There is a clear tendency for those with cultural and economic capital to utilise this capital in ensuring their voice is heard in formal governance processes and when linking with local government and other agencies.

Negative Social Capital

An understanding of the different faces of social capital and the implications of power differentials, also provides an insight into the dark side or downside of the concept, indeed even Putnam has conceded that social capital can be misused in certain circumstances (Field, 2003; Portes and Landolt, 1996; Putnam, 2000). A key negative outcome of social capital is the manner in which social capital can be used to exclude rather than include. This occurs when communities or groups which have strong ties and networks among themselves – bonding social capital – use these ties to ensure that others outside the group are deprived of access to the networks (Portes and Landolt, 1996; Portes, 1998). This disadvantage can be related to the manner in which social capital is treated as a 'club good' to pursue the interests of the club's or section of society's good, rather than as a 'public good' for the good of society as a whole (Aldridge and Halpern, 2002). There is an inherent tendency for those who will benefit from the network to keep the network as closed as possible (DeFilipis, 2001). Clearly then, residents' associations and neighbourhood action groups may use their social capital in a negative way to exclude what they perceive as either undesirable people or undesirable uses from their neighbourhoods (Briggs, 2004). Putnam recognises this when he points out that among other groups, NIMBY ('not in my backyard') movements, often exploit social capital to achieve ends that are antisocial from a wider perspective (Putnam, 2000, p. 22).

This tendency to Nimbyism among individuals and communities against development pressures in their neighbourhood is one of the challenges to the legitimacy of participatory processes (Vigar and Healey, 2002), which are otherwise widely championed as a means of engaging citizens in the policy making process (Bailey and Peel, 2002; Bryson and Crosby, 1992; Healey, 1996, 1997; Sharp and Connelly, 2002). Other limitations and challenges to such participatory processes relate to the extent to which community groups and organisations can claim legitimacy to participate on behalf of their neighbourhoods. We might ask the question how far do the networks and reciprocity of social capital stretch and whom do they encompass? Curry (2001) poses a number of questions focusing on the extent to which community-based decisions are democratic, accountable and representative. Do they promote the views of the entire community, or just certain notables? Is it possible to identify a common view at all, as agreement on who constitutes the community may be contested? Are the decisions made democratic and are those making the decisions accountable to the wider community? These are thorny questions and yet are crucial to understanding the nature of the role that residents' associations play in their neighbourhoods.

The Institutional Context of Social Capital

A final element in the social capital literature that is of interest to the current chapter is the extent to which the political and institutional context influences the extent to which social capital is activated (Maloney et al., 2000; Lowndes and Wilson, 2001; Pennington and Rydin, 2000). Political institutions, and in particular local governance institutions, have a role in encouraging and sustaining civic vibrancy. This is because the approach of elected local authorities together with non-elected partnerships or other institutions help determine the extent to which communities become mobilized. As Lowndes and Wilson (2001) argue, these institutions can enable and support (or disable and frustrate) the citizenry as they are influential in supporting and recognizing the voluntary and community sector (through funding or in their formal recognition of certain groups), in providing the opportunities for participation, in their responsiveness to citizens in decision making and in the extent to which decision making is actually transparent and adequately balances the diversity of different community demands. As Maloney et al. (2000) point out, a city's political opportunity structure affects social capital as it provides the openings or incentives for people to undertake collective action.

In an Irish context, the political opportunity structure has changed considerably in recent years. The last decade or more has witnessed an increased interaction between the State and civil society in Ireland, with a radical restructuring of local governance relationships providing a changed institutional context in which residents' associations act. Traditionally, local government in the Republic of Ireland has been characterised by a limited range of functions, an inward looking and bureaucratic culture, and an absence of a local tax base (for example, see the Barrington Committee Report, 1991). However, in common with many advanced capitalist societies, the 1990s was also marked by a growing shift from government to governance with the emergence of a suite of local development partnerships underpinned by the involvement of the community sector (see Walsh, 1998).

Although partnership structures have become the dominant mechanism in terms of policy formulation and implementation, Broaderick (2002) questions the relationship of partnership and participation, which are often seen as synonymous. At best, Walsh (cited in Broaderick, 2002) suggests that partnership redistributes power among a small elite, while participation shares power among a broader constituency. Much less clear is the ability, capacity and opportunities for the more informal networks of residents' groups at a neighbourhood level to participate in local governance both in terms of the opportunities offered by the institutional structures and their own organizational and resource constraints. These groups, for example, may only mobilise on an occasional basis, often as a reaction against a perceived harmful policy or development proposal. For these groups or associations, campaigning, protest or advocacy are perhaps more realistic alternatives for participation than involvement with partnership processes.

One of the few arenas that enables an opportunity for this layer of informal residents' groups and citizens to become involved in urban and environmental management is the statutory physical planning system. Public involvement and participation in the planning system is enshrined under current legislation (the Planning and Development Act, 2000), and relates to both development plan and development control functions of the local authority. While the 2000 Act increased opportunities for participation, it is within the development control arena where most of the public's interaction with the planning system takes place. Individuals, community groups and organisations have the right to make written submissions on planning applications, or to make a Third Party Appeal against the granting of planning permission. The planning process often becomes one of the few areas in an Irish local governance context that offers formal (statutorily defined) opportunities for participation. The options for residents' associations in terms of participation are thus threefold; they may engage with the formal collaborative arenas offered by partnership; they may utilise the official channels of the formal planning system; or they may engage in the direct and informal 'agonistic' action identified by Hillier (2002), which in effect bypasses more formal channels of interaction. Each option requires differing amounts of bonding and bridging social capital and the nature of the engagement may depend on the underlying cultural and economic capital of the individuals involved. The remainder of this chapter explores these issues in a Dublin context.

Residents' Associations in Dublin

The research on which this chapter is based is an ongoing study of residents' associations in the Dublin area. The role of residents' associations is being explored in six areas chosen on the basis of socio-economic profile, the location and scale of the neighbourhood, the timing of neighbourhood establishment and the level of recent growth or decline. Thus, the study investigates residents' associations in two well-established inner suburban areas, in two outer suburban areas, and in two rapidly developing edge city suburbs.[1] In the current chapter the experiences of residents' associations in the two inner suburban areas are outlined. The two areas in question are contrasting in terms of their socio-economic profile and their recent demographic change, which is summarised in Table 12.1 below.

Case Study Area 1: Kilmainham / Inchicore

The area of Kilmainham/Inchicore is a historic and diverse one. Housing development in the area dates mostly from the late 19th and 20th centuries, with significant amounts of both local authority housing (including Bulfin, Tyrconnell, Bluebell estates) and flat complexes (including Tyrone Place, Islandbridge and St. Michael's Estate). Given its proximity to the city centre the area is undergoing considerable gentrification, with young affluent buyers purchasing the older 19th and early 20th century stock, tenant purchased local authority housing, or occupying new infill housing developments and apartment complexes. In the

Table 12.1 Profile of chosen areas

Location	*Area*	*Income*	*Demographic Change between 1996-2002*
Established Inner Suburb	Ballsbridge/ Sandymount/Nutley	High Income	Low Growth of 3.1 % Most significant growth in the Ballsbridge area (Pembroke East E 11.3%).
Established Inner Suburb	Kilmainham/ Inchicore	Lower-Middle Income	Mixed Growth Rate of 12.4% High rate of growth in the areas closest to the city centre (99.8% growth in Ushers A, 20.4 % Ushers F). Decline in the western area in Inchicore A decline of -5.2% and in Kilmainham A, of -4%

Kilmainham/Inchicore area the issues on which residents' associations focus, are influenced by the social problems in the area, together with the pressures arising from development, redevelopment and renewal.

Case Study Area 2: Sandymount/Ballsbridge/Merrion

The Sandymount/Ballsbridge/Merrion area is located to the South of Dublin City. It is a high-income area, with a relatively stable population and slow growth. The housing stock in the area consists of Georgian, Victorian and Edwardian housing, mixed with housing estate developments dating from the 1940s on. There is very little social housing in the area, and the small amount that did exist in the vicinity of Bath Avenue, has been privatised. It has been described by residents as an established, relatively settled area. New development in the area tends to be confined to former institutional land and subdivision of large residential plots. The main concern of these residents' associations is to mitigate the effects of development in their area. There is a strong emphasis on the protection of both the residential character of the area and unique environmental features existing in the area, notably Sandymount Strand and Dublin Bay. There is also a focus on creating a safe environment through controlling speed, parking and vandalism and monitoring and reporting crime.

In each of the areas the chairperson or secretary of the various residents' association, which had been identified from a wider scoping exercise were contacted by letter and where contact details were available, by phone. Once these individuals were contacted using snowball sampling they were asked to provide the researchers with names of other residents' association committee members. In-depth tape recorded interviews were carried out with the officers (Chairperson, Secretary, Treasurer) and some committee members from each of the residents' associations. The interview schedule used covered the issues of the role of the residents' association, the activities with which they were involved, their links with the wider community and with other actors and agencies. In total thirty interviews have been carried out with residents representing eight residents' associations in the two areas.

Residents' Association, Representation and Bonding Social Capital

While residents' associations themselves are indicative of associational activity (in that they bring people together to co-operate and are thus one factor used in measuring social capital) our research shows that the number of active members is small, usually only 10-15 people, often with an even smaller number of officers responsible for undertaking most of the work. The research revealed that the wider impacts of residents' associations in terms of generating a sense of neighbourliness, reciprocity and trust in the wider neighbourhood community was quite small. The main benefit of the residents' associations in these two areas is for undertaking collective action.

In both of the two inner suburban areas the residents' associations were all founded in response to a perceived need to act in a collective manner. For the most part the impetus was the need to provide a concerted voice in protesting against a threat to their area. In only one of the residents' associations interviewed (a residents' association in a former local authority housing owned estate in the Inchicore area), was the initial stimulus for the set up of the residents' association to provide a community facility – in this instance the provision of a community centre. As a result the emphasis in this residents' association has been the running and upkeep of this centre. This association has also developed a number of other community development initiatives such as a summer project for children. The residents' association and community centre also provided a focal point for the community, particularly for elderly residents. It was important for residents' association committee members in this area that the elderly residents knew that:

> there is support available up here [in the community hall], if there is a problem then they can come up, they mightn't always come up, but they know that they can come up, and you know somebody will act on their behalf (Interviewee 28).

This was the only residents' association where the organisation was seen as being a crucial element in creating a sense of community and neighbourliness and where there was evidence of networks of support and reciprocity that were supported by the residents' association.

The other residents' associations interviewed, including others in the Inchicore area, had been established in response to perceived broader threats to the area and their on-going role was largely focused on maintaining the quality of the physical environment. In general the residents' associations studied are largely reactive and in many cases are at their strongest when they are acting in opposition. As one residents' association committee member pointed out:

> They never seem to be for anything they seem to be against everything so that's why it [residents' association] was set up, basically to stop development (Interviewee 24).

For many of our interviewees the key rationale for the residents' association was the ability to act collectively on behalf of the neighbourhood and give the people of the area a collective voice. There was a sense that a residents' association provided greater power in either lobbying for some facility or service, or lobbying against an unwanted development in the area.

> It gives power to the people and acts as a mouthpiece for the neighbourhood, it gets things across (Interviewee 10).

> When people get together and raise their concerns, they are encouraged when others share those concerns, it gives them more courage (Interviewee 19).

In particular, the benefits of having a residents' association was the extent to which it provided a means of presenting the views of the area to the local authority.

> In the sense that you have a much stronger leverage with local authorities... you can represent your views as being a view from the area. They [the local authority] are happier, as well, dealing with that, rather than having to deal with a large number of people and not being sure they are getting a general view...It also provides a forum for correspondence and for interaction with all different parties and with local county councillors (Interviewee 25).

While often there is a claim to speak on behalf of their neighbourhoods, the residents' associations themselves recognised the limitations of their representativeness. This stemmed from the small numbers of active volunteers and the difficulty of ensuring the wider membership was kept informed of actions being taken on their behalf by the residents' association. The number of active volunteers is a problem shared by almost all voluntary organisations. All of our interviewees believed that it was difficult to ensure the involvement of wider numbers of residents in the committee and the work of the residents' association. The reasons outlined for this lack of participation were in general the time constraints, particularly for residents with young families.

> I think a lot of people are very busy. Young people with both parents working and young children - they probably haven't time - for a lot of people they just want to get on with their lives (Interviewee 13).

The limitation on the number of active volunteers poses problems for some of the residents' associations. The failure to ensure wider participation leaves residents' associations vulnerable, with the potential to become dormant when key individuals step aside or lose interest. This was clearly the case in one of the case study areas where there was almost a sense of despair that it was the 'same few people' that turned up to annual general meetings or got involved in other ways.

The difficulty of ensuring that there are appropriate channels of information flow between the residents' association and the wider community in the area was also explored. All of the residents' associations interviewed had made various attempts to involve and inform the wider community of the issues which the residents' association dealt with. This was usually through a number of fora, notably the Annual General Meeting, the organization of public meetings on key issues and the production of a newsletter. The difficulty for most residents' associations in producing a newsletter was that this was a big commitment and involved having both the expertise to produce a newsletter and the ability to ensure the distribution of the finished product:

> Well I think its difficult - residents' associations, and ours is no exception, are all voluntary, so it's very difficult. We do a newsletter - and its quite informative, but you know it can't cover everything. There's a lot of things going on and it is part of the process of communicating with people in the area about things that are relevant to the

> houses and the road and, that sort of thing, but it's not ideal. …I think it certainly could be improved, but it will always be haphazard and it will be dependent on the quality of the people involved, particular the chairman - and the secretary (Interviewee 21).

This haphazard nature of communication was evident in all of the associations, with many residents bemoaning the difficulty of producing newsletters on a frequent basis and of keeping residents informed. The evidence from the interviews suggests that the flow of information from residents' associations to the wider community is relatively ad-hoc.

Given that much activity is confined to a few individuals and with ad hoc attempts at communication, there is the potential for residents' associations to fail to be representative of the wider community. Likewise, without strong community networks and channels for feedback, it is difficult for the wider body of residents to hold the committee members of the residents' association accountable for their actions (Taylor, 2000).

Bridging Social Capital

On a day-to-day basis the residents' associations interviewed were not a key forum for creating networks of reciprocity and neighbourliness in their communities, however we will reveal later how effective they are at creating bonding capital in times of perceived threat. Their main activities were in engaging with outside agencies on behalf of the wider community and to the best of their ability they acted in the interests of their communities. Thus, the residents' associations in the two areas studied are good at creating bridging social capital (Briggs, 1998; Putnam, 2000) and acting as a node in linking their neighbourhoods with external actors and agencies. Residents' associations are used as the point of engagement with local governance structures, most frequently the local authority but also other governance agencies (partnership and other agencies with a remit for certain areas[2]). This linkage is important, as it can be instrumental in the manner in which local areas can access resources, particularly resources related to environmental improvements. This linkage is also more important in relation to the manner in which neighbourhoods engage with the planning system. In the more affluent area of Nutley in the Sandymount/Ballsbridge/Merrion case study area the underlying cultural and economic capital of the committee members was drawn on in dealing with the planning system and with other elements of local government. Professionals such as solicitors, accountants and engineers were viewed by other committee members as having key skills that were useful in drafting submissions on planning applications and in engaging with officials and councillors in local government.

> Well I think as far as the association is concerned - its been going for a great number of years and it has been very lucky to have had people like solicitors or accountants, surveyors and architects and those kind of people with a bit of expertise. What you don't want is people who join and get on the committee just because they're busybodies. You always need a very strong chairman. The chairman has to be very well selected (Interviewee 22).

In the Bath Avenue Residents' association (also located in the affluent Sandymount/Ballsbridge/Merrion case study area, but with a mixed population), the residents' association has effectively linked the neighbourhood into wider networks of expertise consisting of barristers, engineers and architects from outside the area. The residents' association draw on these experts to give advice on planning and environmental issues:

> Another thing that we've done is put together a very good technical team, which we never had up until about five years ago, a very good technical team that can advise us and help us. Because a lot of the documentation that comes out from different government departments is from teams - now we've ours as well. We have people from, well outside, that will give their services voluntarily to us and then we share this with the other people in the area (Interviewee 8).

In this way the residents' association's role as a bridge to external expertise is extremely important and beneficial to the community at large as is intimated in the quote above.

Residents' associations in the Inchicore area, which is the less affluent location, while not necessarily themselves possessing the social and cultural capital of those in more affluent areas and thus with less expertise at their disposal, also tended to tap into the wider social networks of individual members. This is bridging social capital, which ultimately results in benefits for the community. It was pointed out on a number of occasions that certain individuals played a key role in accessing advice, information or resources:

> Our former chairman x, he was great at building links (Interviewee 12).

> I get advice through friends who work in areas of the civil service, who will sit down with me and give me information (Interviewee 17).

There is thus an interesting contrast between the residents' associations in areas where the residents possessed significant cultural and economic capital which was utilised for the benefit of the entire community, whereas in areas where residents own cultural and economic capital was less, the creation of bridging social capital was an important facet of the residents' association's role. As one of the interviewees from the Sandymount and Merrion residents' association pointed out, many of the residents in the Sandymount area could rely on their own social networks and contacts when dealing with issues affecting their environment, therefore they did not need to engage in the type of collective action which the interviewee saw as being more prevalent in less affluent areas:

> I think like in area x there are a lot more people and in area y there are a lot more people prepared to get dirty for their community than those in Sandymount, because they are all very busy and they all have jobs and they all have nice cars and nice suits and don't really want to go and chain themselves to the Dart[train] line for something. There's probably quite a lot of people who lodge their own objection straight to someone whom they know. They probably know the city Manager personally or this type of stuff. You know there's quite a lot of people living around here who probably have a lot of clout you know, who we should bring on to our committee (Interviewee 5).

The difference in the type of collective action undertaken by residents associations is best exemplified by the engagement and protest over unwanted development in the Inchicore area and in the Nutley area of the Merrion. The protest in both areas is illustrative of the manner in which bonding and bridging social capital is utilized in a negative manner to protect the interests of the existing community and is outlined below.

Engagement and Action, Negative Social Capital?

In situations where there was a perceived threat to their neighbourhood in the shape of unwanted developments the residents' associations were at their most effective, and the type of bonding capital necessary to allow collective action was created. Perhaps it should come as no surprise, given the insights from the literature, that this collective strength and bonding social capital was most in evidence when the residents' associations were acting in an exclusionary manner to protest against certain uses being located in their areas. This portrays some of the negative facets of social capital illustrating the manner in which a residents' association can galvanise significant support when it is believed that an undesirable development will impact negatively on their area.

In Inchicore, a juvenile detention centre, and in Nutley, the use of a former religious retreat house as a residential unit for asylum seekers, were the focus of considerable local opposition. The form the opposition took is illustrative of the differing levels of economic and cultural capital and how these are expressed in social capital and in differing forms of collective action. In both cases the communities were opposed to the siting of these facilities in their communities and appealed to authorities outside their neighbourhoods to intervene to ensure these developments did not go ahead. The processes involved are outlined below.

Broc House was a former residential hostel for students owned by the Franciscan order. The Commissioners for Public Works bought it in 2000 at a cost of €9.2 million with the aim of using the building as a reception centre for asylum seekers. At the time of the purchase there was a crisis over lack of accommodation for asylum seekers and the Department of Justice was actively looking for accommodation. It was believed that such a use could be accommodated on the site as the planning authority for the area, Dublin city council, had previously stated that the change of use from a 'residential hostel for student accommodation' to 'residential hostel and hotel' was exempted development.

There was considerable concern among local residents when it became apparent that Broc House was to be used as a reception centre for asylum seekers. While the residents were guarded in the manner in which they expressed their opposition to the proposed development, it is evident that many residents were irrationally fearful and concerned about potential threats to their property. Interviewees expressed fears that the potential occupiers of the building might be 'petty thieves', or that there would be problems with unruly behaviour, particularly at night. There was a clear awareness of the sensitivity of this issue among those interviewed and it was stressed by a number of interviewees that the residents were not opposed to asylum seekers *per se*, but that they were concerned with the ability of the facility to accommodate the numbers of people envisaged. Such an approach of expressing opposition in terms of rational and objective concerns about the effects of unwanted development has been identified by Dear (1992) as the second stage in NIMBY opposition. However, fear of 'otherness' certainly underlay much of the opposition (Sibley, 1995). One of the issues raised by the residents' association in relation to the use of Broc house as a reception center for asylum seekers was that the infrastructure in the area was insufficient for asylum seekers. This is in direct contrast to their own assessment of the facilities and services in the area, which they themselves believed were good!

Initial concerted action in relation to the proposal to house asylum seekers in Broc House took the form of the residents' association meeting themselves to discuss the issue and subsequently organizing meetings between the local residents and the Department of Justice on a number of occasions. These meetings did nothing to assuage the residents' fears regarding the numbers to be housed, the management of the center and potential security issues. As it had been established that the change of use proposed was exempted development, and that no planning permission was required, the potential to voice these concerns within the planning process did not exist. Judicial review proceedings were the only, formal and expensive option for challenge available to the local community. Given the potential costs involved in such proceedings, and the sensitivity of the issue (there were clearly residents in the area who were very uncomfortable with the idea of opposing such a use), the residents' association drew back from the process at this point and a subcommittee of the association comprising those residents living in close proximity to Broc House was formed. It was this sub-committee, which was separate to, but supported by the residents' association, which eventually took judicial review proceedings against the development. The residents' association had thus created a platform for people to come together to discuss the issue (creating bonding social capital) and then supported the subcommittee by contributing in a small way to the costs of taking these proceedings. It was reiterated by the residents' association that this challenge, and the residents' association support of this challenge was based on legal grounds:

> We have made a contribution to the fund that the local residents have raised to legally fight the case and by the way the legal issue was nothing to do with asylum seekers, it was to do with the use of the building (Interviewee 21).

This course of action is an expensive one, requiring a legal team and the potential that costs may be awarded against those taking the judicial review proceedings. The ability of the residents in the area to take such proceedings is indicative of the financial capital available to them as a group. In May 2004 the residents' objection was overruled, allowing the State to use the facility for the purposes of accommodating asylum seekers. However, given the reduction in applicants seeking asylum, the need for the facility is no longer so pressing. Thus, in effect the residents' actions have effectively blocked the use of the facility for these purposes (for over 3 years). The residents' opposition has been powerful and firmly embedded in the official legal channels. This was a professional and well-resourced protest, but also a protest that was conscious of the political sensitivities of the issue and motivated by a desire to ensure that opposition was seen as 'about uses'. The residents' association did not want to be seen to be acting against the public interest:

> We didn't make any headlines in the papers at all, we didn't express any opinions on the issue because it was, at the time, a very delicate matter and we would not like to be held up as acting against the public interest (Interviewee 23).

The arguments against the facility were thus cloaked in the technocratic and legal discourse of the judicial review and are thus removed from the messy and vocal arena of public protest, indeed from the quotation above it is clear that the residents avoided heavy publicity. There was no evidence of open political lobbying, but this may have occurred behind the scenes. Such an approach to protest was facilitated by the available financial and cultural capital of the individual residents, which was initially brought together in the forum of the residents' association.

This approach contrasts with that which pertained in the Inchicore case study area, where there was an organized campaign of protest against another human services facility – in this case a juvenile detention centre for young offenders. In 2001 under the provisions of the Children's Bill which was then under consideration, the Department of Justice, Equality and Law Reform, in conjunction with the prison service, were required to provide separate detention facilities for juvenile offenders. A search for suitable sites identified a number of possible locations in the Dublin area, one of which was an Office of Public Works owned site on Jamestown Road in Inchicore. The local community were strongly opposed to the proposed siting of such a facility in their area and the residents' association in Inchicore spearheaded a concerted campaign of action against the location of the facility at this site. In contrast to the low-key nature of the action against the facility in Nutley, the situation in Inchicore involved much more active protest.

Unlike the situation in Nutley the residents' association felt ill equipped to deal with the issue, expressing concerns that they felt they were "not the most able individuals from the area", or the "most literate", but there was a sense that there was strength among the residents who were "Dublin working class people". The contrast was made with the protest against a major office development in Kilmainham, where the residents had better organisational resources and had established their own web site, but were less successful in engaging in direct action:

> There seemed to be more of a gutsy reaction here, even though they [the residents' association in Kilmainham] seemed to be more able in what they were doing, it wasn't having the same impact. We went down to help with the road protest and not too many people were there and it was a pity. It was as if people were a bit slow to react to it. Whereas we blocked the road here when we had a road protest along Tyrconnnell road (Interviewee 33).

In their campaign against the proposed detention centre the Inchicore residents' association galvanised support against the proposed development from the entire Inchicore area. This was achieved by presenting the proposed development as a threat to the whole area of Inchicore, rather than simply a threat to those living in close proximity to the site. This storyline was shaped by focusing on the impact that visitors to the detention centre would have when they would congregate in the area before and after visiting hours.

The community action against the proposal was galvanised by knocking on doors, by organising public meetings and by lobbying local and national political representatives for the area. The residents' association acted to build bridging social capital in a number of ways. They sought the support of the political parties, particularly the opposition politicians for their area. They also sought advice from outside sources, this included raising funds to obtain legal advice from a solicitor, and using the committee's social networks, for example they sourced information from the Gardai (police) and Prison officers, who were friends of committee members, on the likely impacts of such a facility. In presenting their opposition to an external audience the residents' association focused on the point that Inchicore and the surrounding area already had an undue concentration of social problems, and that it would be unfair to site a further undesirable development in the area.

In contrast to the situation in the more affluent area, the residents' association sought out publicity and went on radio to talk about the issue and organised street protests including a protest outside the Department of Justice, Equality and Law reform. The protest and opposition to the development took the form of much more informal direct action, lobbying, marches, public meetings and engagement with the media (Hillier, 2002) than was the case in Nutley. This required a greater amount of collective action and mobilisation of the community. In effect a degree of bonding social capital needed to be activated by the residents' association in response to the perceived threat. In this instance the formal arena of the planning system was not available to the community as the Department of Justice, Equality and Law reform were only evaluating sites, although the residents' legal advice was that it would be on planning grounds that the development would succeed or fail. Eventually, the site was deemed unsuitable for the proposed development and the residents' association believed that this was as a result of the limited access points to the site and other possible planning problems.

In both of the examples outlined the aim was to block what were perceived as undesirable uses locating in their neighbourhoods. In both areas bonding social capital was created which was fundamentally exclusionary, but the nature of the collective action that resulted was different in each case. In Nutley, the nature of the cultural and economic capital available to residents was such that the residents' voices would be clearly heard in the legal system. A small number of affluent well-connected individuals were all that was required. There was no need to depend on direct action; indeed the tendency was to play down the opposition, rather than to draw attention to it. In Inchicore the residents' association had a greater compulsion to mobilise opposition to the proposed development. In this case the amount of opposition was important given the inability to participate in a more formal planning or legal arena. In this case the residents' association had to build up bonding social capital in order to ensure collective action, whereas this was less important in the Nutley case, where strength in numbers was not important once a judicial review had been applied for. In Inchicore, the residents' association was a useful forum for bridging and linking to external expertise, with the social networks of committee members being used to access information from external sources. In Nutley there was less need to rely on external expertise, as considerable expertise existed among the residents' body.

Conclusion

The research outlined in this chapter has explored the nature of the role and activities of residents' associations in the context of an understanding of the concept of social capital. The research documented above has shown that only in a few cases have residents' associations helped create the positive, warm elements of social capital by establishing networks of reciprocity and social support and providing community infrastructure. In this example there are clear positive impacts for the wider community who effectively free ride on the activities of a few. What the research shows is the downside of social capital, the exclusionary nature of a bonded community and the extent to which an external threat is a catalyst to overcoming collective action problems. In general residents' associations are much more effective at creating bonding social capital when there is a perceived threat which propels residents together to act collectively in a defensive manner. In times of external threats the residents' associations studied were also good at creating bridging social capital, linking their communities into networks of expertise and advice. Even in the absence of major threats they performed the role as a link to external bodies – particularly local authorities.

In a particularly affluent area, the extent to which the residents own cultural capital can underlie their social capital and can shape the nature of collective action has been highlighted. The financial capital available to the residents in Nutley meant that they could utilise the more powerful, legalistic and technocratic forum of the courts to mount their challenge. However, while the research supports Bourdieu's contention of the importance of underlying cultural and economic capital somewhat, it does illustrate as Field (2003) has pointed out that social capital is not purely something which is confined to the better off, and that in any area people can utilise their social networks in a manner that protects their interest. In Inchicore the residents' association utilised their social capital to obtain their desired outcome, in the absence of significant amounts of cultural or financial capital. What is revealed by the research is that their protest required much greater amounts of human capital in order to mount a convincing and effective challenge. This shows that there can be trade offs in the way in which different forms of capital are utilised. In conclusion, while social capital is a useful concept, the nuances of how it is created and applied require further thought, if this kind of research is to contribute meaningfully to the debate of the future of towns and cities in Ireland.

Acknowledgements

The authors would like to acknowledge the support of the Royal Irish Academy's Third Sector Research Programme for funding the research on Residents' Associations, Neighbourhood and Community Development from which the research outlined in this paper is drawn.

Notes

[1] Sandymount/Ballsbridge/Merrion, Kilmainham/Inchicore, Glasthule/ Sandycove, Finglas, Rush/Skerries, Lucan.

[2] In the case study areas residents' associations engaged with and were represented on a number of these agencies. In Inchicore the Bulfin Residents association was represented on the St. Michael's Task Force and in Sandymount/Ballsbridge/Merrion, the Bath Avenue and District Residents'

Association was represented on the Council of the Dublin Docklands Development Authority.

References

Aldridge, S. and Halpern, D. (2002) *Social Capital: A discussion paper, Performance and Innovation Unit*, London: UK Cabinet Office. http://www.piu.gov.uk/2001/futures/socialcapital.pdf.

Bailey, N. and Peel, D. (2002) 'Building Sustainable Networks: A Study of Public Participation and Social Capital', in Rydin. Y. and Thornley, A. (eds) *Planning in the UK: Agendas for the New Millennium*, Aldershot: Ashgate.

Barrington Committee (1991) *Local Government Reorganisation and Reform – Report of the Advisory Expert Committee*, Dublin: Government Publications.

Briggs, X. de Souza, (1998) 'Brown kids in white suburbs: Housing mobility and the many faces of social capital', *Housing Policy Debate*, Vol. 9, pp. 177-221.

Briggs, X. de Souza, (2004) 'Social Capital: Easy Beauty or Meaningful Resource?', *Journal of the American Planning Association*, Vol. 70, pp. 151-158.

Bryson, J. and Crosby, B. (1992) *Leadership for the Common Good: Tackling Public Problems in a Shared Power World*, San Francisco: Jossey Bass.

Butler, T. and Robson, G., (2001) 'Social Capital, Gentrification and Neighbourhood Change in London: A Comparison of Three South London Neighbourhoods', *Urban Studies*, Vol. 12, pp. 2145-2162.

Bourdieu, P. (1986) 'The Forms of Capital', in Richardson, J.G. (ed.) *Handbook of Theory and Research for the Sociology of Education*, New York: Greenwood Press, pp. 241-258.

Broaderick, S. (2002) 'Community development in Ireland – a policy review', *Community Development Journal*, Vol. 37, pp. 101-110.

Coleman, J. S. (1988) 'Social Capital in the creation of human capital', *American Journal of Sociology*, Vol. 94 Supplement, pp. s95-s120.

Curry, N. (2001) 'Community Participation and Rural Policy: Representativeness in the Development of Millennium Greens', *Journal of Environmental Planning and Management*, Vol. 44, pp. 561-576.

Dear, M. (1992) 'Understanding and Overcoming the NIMBY Syndrome', *Journal of the American Planning Association*, Vol. 58, pp. 288-300.

De Filippis, J. (2001) 'The Myth of Social Capital in Community Development', *Housing Policy Debate*, Vol. 12, pp. 781-806.

Democracy Commission (2004) *Disempowered and Disillusioned but not Disengaged – Democracy in Ireland: A Progress Report*, Dublin: Democracy Commission.

Field, J. (2003) *Social Capital*, London: Routledge.

Forrest, R., and Kearns, A. (2001) 'Social Cohesion, Social Capital and Neighbourhood', *Urban Studies*, Vol. 38(12), pp. 2125-2143.

Government of Ireland (2000) *White Paper on a Framework for Supporting Voluntary Activity and for Developing the Relationship between the Community and Voluntary Sector*, Dublin: Stationery Office.

Granovetter, M. (1973) 'The strength of weak ties', *American Journal of Sociology*, Vol. 78(6), pp. 1360-1380.

Granovetter, M. (1995) *Getting a Job: A Study of contacts and careers*, 2nd edn, Chicago: University of Chicago Press.

Hall, P. (1999) 'Social Capital in Britain', *British Journal of Political Science*, Vol. 29, pp. 417-461.

Healey, P. (1996) 'Consensus-building Across Difficult Divisions: New Approaches to Collaborative Strategy Making', *Planning Practice and Research*, Vol. 11, pp. 207-216.

Healey, P. (1997) *Collaborative Planning, Shaping Places in Fragmented Societies*, London: Macmillan.

Healy, T. (2004) *Social Capital: Some Policy and Research Implications for New Zealand*, Unpublished Paper.

Hillier, J. (2002) 'Direct Action and Agonism in Democratic Planning Practice', in Allmendinger, P. and Tewdwr–Jones, M. (eds) *Planning Futures: New Directions in Planning Theory*, London: Routledge, pp. 110-135.

Kearns, A. and Parkinson, M. (2001) 'The Significance of Neighbourhood', *Urban Studies*, Vol. 38, pp. 2103-2110.

Larsen L., Harlan, S.L., Bolin, B., Hackett, E. J., Hope D., Kirby A., Nelson A., Rex T. R., and Wolf S. (2004) 'Bonding and Bridging Understanding the Relationship between Social Capital and Civic Action', *Journal of Planning Education and Research*, Vol. 24, pp. 64-77.

Lowndes, V. and Wilson, D. (2001) 'Social Capital and Local Governance: Exploring the Institutional Design Variable', *Political Studies*, Vol. 49, pp. 629-647.

Maloney, W., Smith, G. and Stoker, G. (2000) 'Social Capital and Urban Governance: Adding a More Contextualized "Top Down" Perspective', *Political Studies*, Vol. 48, pp. 802-820.

Mohan, G. and Mohan, J. (2002) 'Placing Social Capital', *Progress in Human Geography*, Vol. 26, pp. 191-210.

Office for National Statistics, Social Analysis and Reporting Division, (2001) *Social Capital: A review of the literature*, October 2001.

OECD, (2001) *The Well-being of Nations: The Role of Human and Social Capital*, Paris: OECD.

Pennington, M. and Rydin, Y., (2000) 'Researching Social Capital in Local Environmental Policy Contexts', *Policy and Politics*, Vol. 28, pp. 233-249.

Portes, A. (1998) 'Social capital: its origins and applications in modern sociology', *Annual Review of Sociology*, Vol. 24, pp. 1-24.

Portes, A. and Landolt, P. (1996) 'The Downside of Social Capital', *The American Prospect*, Issue 26, pp. 18-21.

Purdue, D. (2001) 'Neighbourhood Governance: Leadership, Trust and Social Capital', *Urban Studies*, Vol. 38, pp. 2211-2224.

Putnam R. (1993) *Making democracy work: civic traditions in modern Italy*, Princeton: Princeton University Press.

Putnam, R. (2000) *Bowling Alone: The collapse and revival of American community*, New York: Simon and Schuster.

Sharp, L. and Connelly, S. (2002) 'Theorising Participation: Pulling Down the Ladder', in Rydin, Y. and Thornley, A. (eds) *Planning in the UK: Agendas for the New Millennium*, Aldershot: Ashgate.

Sibley, D. (1995) *Geographies of Exclusion*, London: Routledge.

Taylor, M. (2000) 'Communities in the lead: Power, organisational capacity and social capital', *Urban Studies*, Vol. 37, pp. 1019-1035.

Taylor M. (2003) *Public Policy in the Community*, Basingstoke: Palgrave Macmillan.

Vigar, G. and Healey, P. (2002) 'Developing Environmentally Respectful Policy Programmes: Five Key Principles', *Journal of Environmental Planning and Management*, Vol. 45, pp. 517-532.

Walsh, J. (1998) 'Local Development and Local Government in the Republic of Ireland: From Fragmentation to Integration?', *Local Economy*, Vol. 12, pp. 329-341.

Woolcock, M. (1998) 'Social capital and economic development: Toward a theoretical synthesis and policy framework', *Theory and society*, Vol. 27, pp. 151-208.

Woolcock, M. and Narayan, D. (2000) 'Social Capital: Implications for Development Theory, Research and Policy', *The World Bank Research Observer*, Vol. 15, pp. 225-249.

Chapter 13

Housing Policy, Homeownership and the Provision of Affordable Housing

Declan Redmond and Gillian Kernan

Introduction

In Ireland, the promotion of home ownership has been evident for many years. Polices such as the abolition of residential rates in 1978, the abolition in 1994 of a residential property tax which had been introduced in 1984, the absence of capital gains tax on the sale of the households principal residence, no stamp duty on new houses for owner occupiers, interest relief on mortgage repayments, a first-time buyers grant (which was abolished in 2002), a generous tenant purchase scheme for local authority tenants, grants offered to local authority tenants for the surrender of their dwelling and moving to owner occupation, all testify to the centrality of promoting owner occupation over the past thirty years (National Economic and Social Council, 1988; Downey, 2003). As Table 13.1 shows, collectively these policies have aided the expansion of homeownership to the point where almost 80 per cent of households own their own home, one of the highest rates in the European Union (European Union, 2002). The encouragement of owner occupation has been seen as a political and social priority by successive governments, and while there may have been differing emphases on the degree to which this should be promoted and conversely, the degree to which other housing tenures should be encouraged, no government in the past 30 years has seriously articulated, still less promoted, an alternative to the domination of home ownership. As a result, social rented housing and private renting have very much taken a secondary role in housing provision and housing policy (Drudy and Punch, 2001, 2002).

It is therefore not surprising that in the past decade, when house prices have escalated at an astonishing rate and given rise to all manner of access and affordability problems, that the policy responses have still remained fixated on access to owner occupation (Bacon and Associates, 1998, 1999, 2000; Memery, 2001; Downey, 2003). The aim of government housing policy is "to enable every household to have available an affordable dwelling of good quality, suited to its needs, in a good environment and, as far as possible, at the tenure of its choice" (Department of the Environment, Heritage and Local Government, 2004). On the face of it this is an admirable set of policy objectives, emphasising access and

Table 13.1 Dwellings by type of tenure in Ireland

Tenure	1991		2002	
	N	per cent	N	per cent
Owner occupied - with mortgage	355,851	34.9	484,774	38
Owner occupied - outright	387,278	38.0	461,166	36
Being purchased from local authority	65,256	6.4	44,783	3
Rented from local authority	98,929	9.7	88,206	7
Rented unfurnished	18,094	1.8	25,883	2
Rented -furnished	63,330	6.2	115,576	9
Free of rent	21,589	2.1	21,560	2
Not stated	9,396	0.9	37,669	3
Total	1,019,723	100	1,279,617	100

Source: Central Statistics Office (1996, 2004a)

affordability ostensibly in a tenure-neutral manner. However, according to the government, "the general principle underpinning the policy approach to the housing objective is that those who can afford to do so should provide for their housing needs either through home ownership or private rented accommodation and that those who will be unable to provide for their own housing needs should have access to social housing" (Department of the Environment, Heritage and Local Government, 2004). In reality, the private provision of owner occupied housing is meant to deliver on the policy goals set. Rather than being a tenure-neutral policy, in fact we have a tenure-led policy. This chapter examines what happens when house prices rise to such an extent that it generates a problem of access to house purchase that is different in scale and nature to any previous market problem. The chapter summarises the growth of the market over the past decade, examines the nature of the affordability problem that the housing boom has created and finally, investigates some of the key policy responses to this crisis of affordability, focusing in particular at the provision of subsidised affordable housing. One of the linguistic indicators of the change in Irish housing has been the invention of the category of 'affordable housing' over the past few years. It is defined in a fairly narrow way to mean the provision of subsidised private housing to aspiring house purchasers who, prior to the housing boom, would have been able to afford a home on the open market.

The Housing Market, Economic Growth and Affordability

While this is not the place to discuss in detail the economic boom of the past decade, it is worth highlighting in very general terms some of the key factors driving the housing market. A number of important factors led to the house price explosion in the Irish housing market over the past decade. On the economic side, low interest rates for entry to the EMU, high increases in employment, and wage and incomes increases from a growing economy put increasing pressure on the demand for housing (Nolan et al., 2000). On the supply side, the response to these shifts in demand were slow to begin with, but began to pick up in the late 1990s. Demographically, natural population increase alongside extensive in-migration from return emigrants and immigration by non-EU nationals has also driven demand heavily. Since the short-run supply of housing is relatively inelastic, the shift in demand due to growth in disposable income, growing employment levels and lower interest rates resulted in prices escalating rapidly. However, supply did respond by the late 1990s. Figure 13.1 plots the total number of dwellings completed over the period 1993 to 2003. The main characteristic of the Irish housing market is that 91 per cent of all houses completed over the past ten years were private houses, while social housing accounted for only 9 per cent of the total completions. Moreover, the rate of house completions in recent years has been the highest per capita in the European Union. One of the more interesting aspects of the recent development boom is that some of the statistical evidence suggests that in the period 1996-2002, as many as 70,000 of the new dwellings completed were second homes (McCarthy, Hughes and Woelger, 2003). This raises the question of

Figure 13.1 New housing completions: State

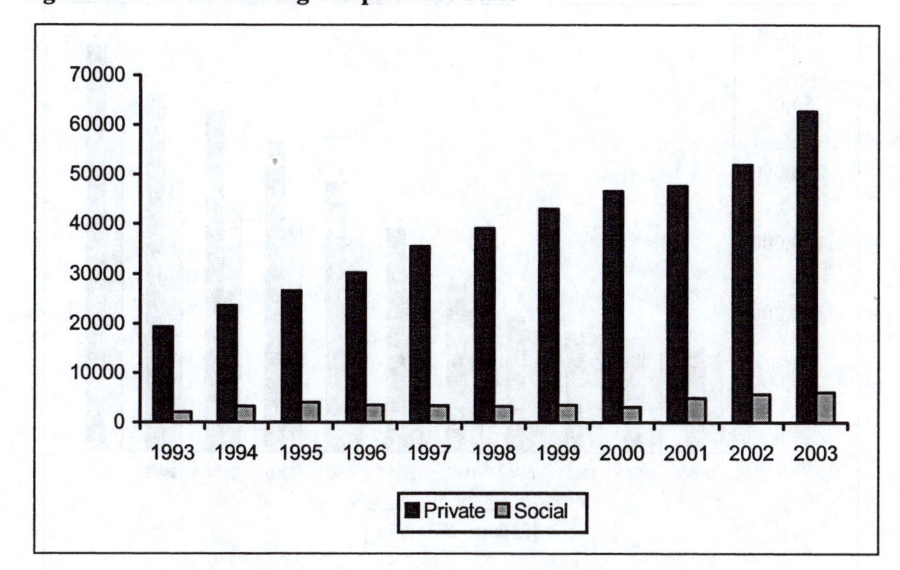

Source: Department of the Environment, Heritage and Local Government (1995-2003)

the precise nature and composition of housing demand and of the efficiency and effectiveness of market provision. Although supply increased very substantially, the level and scale of demand for new housing resulted in extraordinarily rapid house price inflation (Bacon and Associates, 1998, 1999, 2000; Drudy and Punch, 2001, 2002; McNulty, 2003). Figure 13.2 shows changes in new house prices nationally and also in Dublin City for comparison. While the rapid increases in house prices are evident from 1997 onwards, what is also evident is the even more rapid increases in the Dublin area. Over the decade 1993-2003, new house prices increased by 221 per cent nationally while in Dublin new house prices increased by 286 per cent. The second hand market was even more buoyant. Over the same period second hand house prices increased by 296 per cent nationally while in Dublin they increased by an extraordinary 362 per cent. The rate of these price increases, especially when compared to general inflation, suggest the emergence of some kind of affordability problem.

Affordability and Homeownership

The debate about affordability is a very complex one, with a variety of different methods being used to measure affordability. Duffy (2004) distinguishes between affordability measures, which measure access to the market, and those that

Figure 13.2 New house prices in the State and Dublin (€)

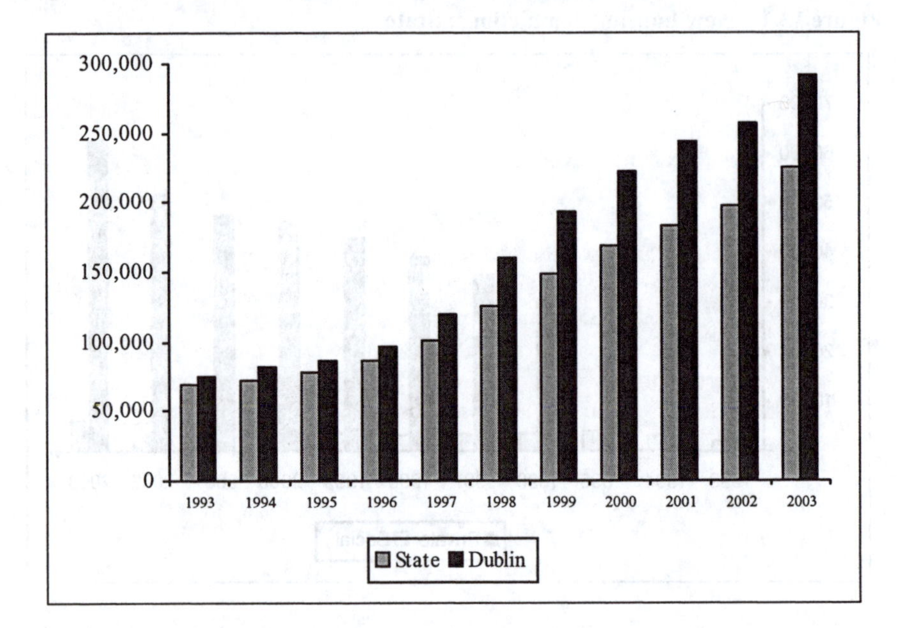

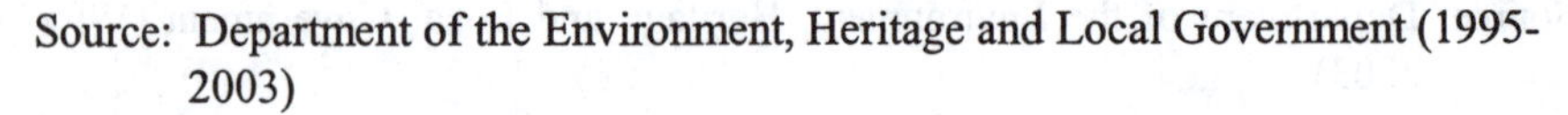
Source: Department of the Environment, Heritage and Local Government (1995-2003)

measure the burden of mortgage repayments for those who already own their own home. For the former he lists the followings methods of measurement: house price to earnings ratios; loan to incomes ratios and deposit to income ratios. For the latter he lists the following: service cost of the mortgage stock; debt-service to income general ratio and housing expenditure to total expenditure ratio. While all of these measures have their advantages and disadvantages, there is an argument about affordability that needs to be highlighted. Affordability is about the relationship between income and housing costs and is in essence about making a judgement regarding what is an appropriate housing cost for a household to bear (Hancock, 1993; Downey, 2003; O' Sullivan and Gibb, 2003). In other words, it is about making a judgement about the residual income available to a household after housing costs have been deducted from household income and stating whether it is sufficient. This judgement is not merely a technical one but one which is in essence social and political. For example, social housing tenants in Ireland generally have a very low, indeed nominal, rent that is very often under 10 per cent of their income (Fahey et al., 2004). However, this low ratio of housing costs to income tells us little or nothing about the sufficiency of their residual income, which may be entirely composed of welfare payments. Conversely, somebody on a high income may pay 40 per cent of their income on housing costs but the residual income may be more than sufficient. Another of the central problems in assessing the extent of the affordability problem, having agreed a measure of affordability, is in attempting to quantify the number of households affected, or as Duffy (2004) puts it, to quantify the number of those who have barriers in accessing house purchase.

In a similar vein, Downey (2003) highlights a number of ways of defining affordability. He argues that for households with above average earnings, affordability can mean a price barrier to entering home ownership that may lead to households putting the purchase of a house on hold and relying on the private rented sector. For households with average earnings and below, the un-affordability of access to private housing can mean that the costs of accessing and consuming housing lead to living in poor housing conditions, to default and living in arrears, to possession, eviction and homelessness, ultimately to a denial to housing. In summation, Downey argues that to be able to 'afford' means to be able to do or spare something without risking financial difficulties or undesirable consequences. So, measuring affordability is both a conceptually and technically difficult task. This has not stopped many self-interested market commentators from arguing that there is no affordability problem. However, the evidence suggests otherwise.

Duffy's (2004) useful definition of affordability, which distinguishes between access to the market and affordability when in the market, provides a useful way of organising a selection of evidence on affordability. And, in that regard, it must be said that the available data on incomes and housing costs in Ireland is, at best, very patchy. With regard to those already paying a mortgage there is some statistical evidence available, although it must be said that it is of limited use.

Table 13.2 examines the relationship between changes in house prices in the State and Dublin against rises in general consumer prices. We can see that overall house prices rose extremely rapidly in the period 1995 to 2003 when compared with consumer prices. In particular, however, we can see the extraordinarily rapid increases in both new and second hand house prices in Dublin. In the space of a year, between 1997 and 1998, house prices rose by an average of 35 per cent in Dublin, as compared with a rise of 2.4 per cent in consumer prices. While the rate of house price increase has decreased, house prices are still increasing rapidly when compared with general inflation. These figures showing the explosion in house prices would suggest that at least some sections of society have problems of affordability. Fahey (2004) has undertaken a comprehensive analysis of the Household Budget Surveys, which are based on samples, to analyse affordability. Figure 13.3 shows the results of the analysis that compares housing costs with total housing expenditure. It shows clearly that there has been a significant rise in housing costs for those in the private rented sector, with rents rising to over 20 per cent of housing expenditure by 2000. For owners with a mortgage, there has been a slight rather than sharp rise while local authority renters have remained fairly stable. However, the trend for owners with a mortgage must be treated with some caution as this includes those who have a mortgage for a number of years as well as recent first time buyers, thus averaging out what are likely to be significant differences.

Table 13.2 House prices and Consumer Price Index

Year	Change in New House Prices-State per cent	Change in Second Hand House Prices-State per cent	Change in New House Prices - Dublin per cent	Change in Second Hand House Prices-Dublin per cent	Change in Consumer Prices per cent
1995	7.2	6.3	5.7	7.5	2.5
1996	11.8	15.2	12.0	17.4	1.6
1997	17.2	20.0	23.4	24.5	1.5
1998	22.6	31.0	34.2	35.6	2.4
1999	18.5	21.4	20.4	19.4	1.6
2000	13.9	16.7	14.6	17.3	5.6
2001	8.1	8.2	9.6	8.5	4.9
2002	8.3	10.5	5.4	11.0	4.6
2003	13.3	16.3	13.9	19.5	3.5

Source: Department of Environment, Heritage and Local Government (1995-2003)

Figure 13.3 Housing costs as a proportion of household expenditure (%)

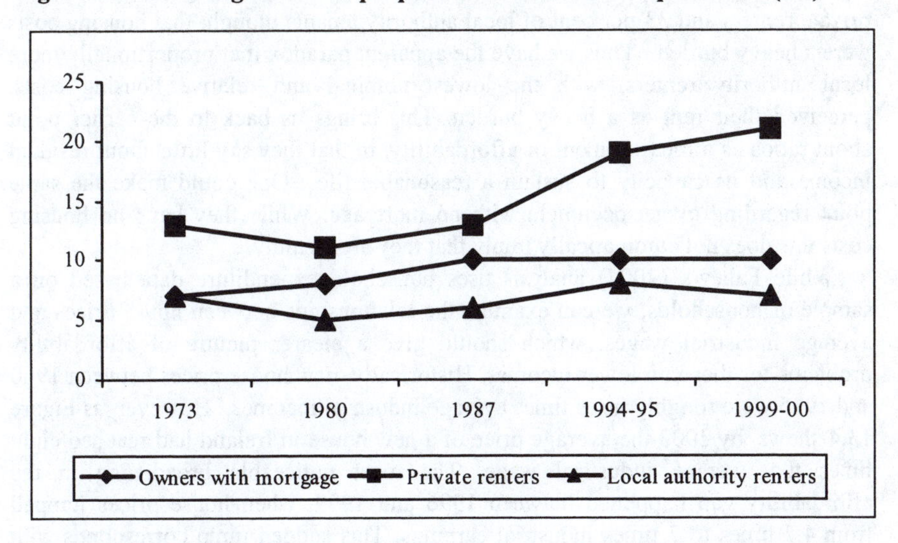

Source: Fahey (2004)

There is a more general warning about these figures and that is that while ratios can tell us something regarding the changing relationship between housing costs and household expenditure, they reveal little *per se* about the burden of housing costs or, conversely, about residual incomes and their adequacy.

More recent figures from the Quarterly National Household Survey on affordability (Central Statistics Office, 2004b) reveal that first time buyers are more likely to have higher mortgages than the average for all owner occupiers with mortgages. For example, of the 240,7000 households who purchased since 1996, 24 per cent had monthly mortgage repayments in excess of €600, while just 3 per cent of those who purchased prior to 1996 had a mortgage of over €600, this being partly attributable to the age of the mortgage and also partly to the increase in house prices since 1996. However, this data set is seriously handicapped by the lack of income or expenditure data which makes it impossible to compare the actual housing costs with income and thus no measurement of affordability can be made. The Irish National Survey of Housing Quality (Watson and Williams, 2003), which was conducted in 2001-2002, undertakes analysis of affordability on the basis of a sample that compares housing costs with net income. Once again, private renters fare worst with 28 per cent paying more than one third of net income on housing costs. Only 6 per cent of owners with a mortgage paid over a third of net income on housing costs while 11 per cent of first time buyers did so. Only 1 per cent of local authority renters paid more than one third of their income on housing costs. However, very interestingly, all the sample households were asked about their perception of the burden of housing costs. Eleven per cent of first time buyers, 13 per cent of owner-occupiers with a mortgage, 20 per cent of private renters and 33 per cent of local authority tenants thought that housing costs were a heavy burden. Thus we have the apparent paradox that proportionally more local authority renters, with the lowest nominal and relative housing costs, perceived their rent as a heavy burden. This brings us back to the earlier point about ratios as a measurement of affordability, in that they say little about residual income and its capacity to sustain a reasonable life. One could make the same point regarding owner-occupiers with no mortgage. While they have no housing costs this does not automatically imply that they are wealthy.

While Fahey's (2004) analysis uses household expenditure data based on a sample of households, we can examine the relationships between house prices and average industrial wages, which should give a clearer picture of affordability problems for those on lower incomes. Historically, new house prices between 1960 and 1995 were roughly three times average industrial incomes. However, as Figure 13.4 shows, by 2003 the average price of a new house in Ireland had reached eight times the average industrial wage. The most noticeable broadening in the affordability gap happened between 1996 and 1999, when house prices jumped from 4.7 times to 7 times industrial earnings. This sudden jump corresponds with the highest increase in house prices over the ten years, when house prices increase by 24 per cent between 1997 and 1998, with only a 4.3 per cent Increase in average industrial earnings. So in that year house

Figure 13.4 Ratio of house prices to average industrial wages (%)

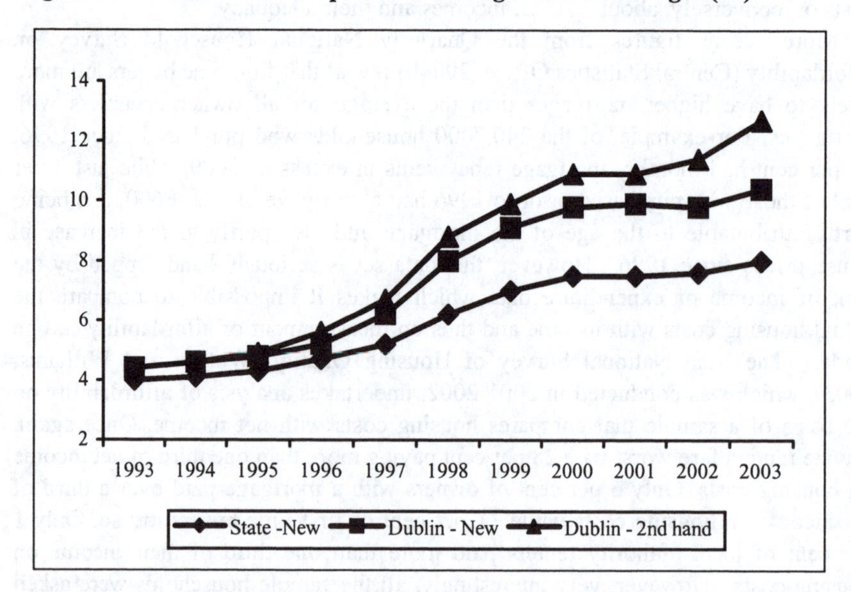

Source: Department of the Environment, Heritage and Local Government (1995-2003)

prices increased at six times the speed of average earnings. The ratio worsens for those attempting to purchase in Dublin. By 2003, new house prices in Dublin were 10 times the average industrial wage and second hand prices were a multiple of 12 times the industrial wage. At the very least, these figures suggest a serious problem of affordability and access to home ownership for those on low wages.

With regard to measuring access to the market, this is perhaps one of the more difficult measures and certainly, with respect to robust evidence, one of the weakest. One of the consequences of the rapid escalation in house prices has been the consequences for other tenures, with evidence of housing need increasing via the demand for local authority accommodation and for subsidised private rented housing. Housing need for local authority accommodation, as measured by the tri-annual Assessments of Housing Need, has increased rapidly since the onset of the housing boom. In 1996 there were 27,427 households in housing need, but by 1999 this had jumped to 39,176 households. By 2002 this had increased again to 48,413 households. There has been an increase of a similar order of magnitude in demand for subsidised private rented housing. The number of households claiming rent supplement, which subsidises the market rent in the private rented sector, has increased from 30,000 to 60,000 between 1994 and 2003 (Department of Social and Family Affairs, 2003). Expenditure from this scheme has increased from €69m in 1995 to €331m in 2003. While there are arguments regarding the robustness of the housing needs methodology, the continued upward trends in housing need assessment and claims for rent supplement, certainly demonstrates a sharp increase in housing need and is evidence of the impact of the problem of affordability.

Further evidence of the impact of house price escalation and attendant affordability problems can be see in the estimates of affordability made by local authorities in their Housing Strategies. Part V of the Planning and Development Act, 2000, requires local authorities to undertake a general assessment of housing need and demand in their area in a Housing Strategy, which is distinct from the assessments solely for local authority housing (Department of the Environment and Local Government, 2000b, 2000c). The Act stipulates that a household has an affordability problem when their income:

> would not be adequate to meet the payments on a mortgage for the purchase of a house to meet his or her accommodation needs because the payments calculated over the course of a year would exceed 35 per cent of that person's annual income net of income tax and pay related social insurance (Government of Ireland, 2000a, p. 109).

While it is unclear as to how the Government chose this particular definition of affordability, its application to the forecast of affordability in the housing strategies of local authorities revealed very high levels of affordability problems in the future. For example, in the Dublin region, on average local authorities predicted that 35 per cent of future households would have an affordability problem based on the above definition (Focus Ireland et al., 2002). While some of this demand would be met by direct local authority provision, some would also be met by private developers and by the provision of subsidised affordable housing.

Affordability and Government Policy Initiatives

As a consequence of the housing boom and problems of access and affordability, a number of analyses of the housing market were undertaken for government and a series of new policy innovations and responses were introduced (Bacon and Associates, 1998, 1999 and 2000; Department of the Environment and Local Government, 1998, 1999, 2000a; Government of Ireland, 2000b). In general terms, a number of fiscal measures were taken which were aimed at managing demand for housing. For example, rates of stamp duty were reduced for first time purchasers of new housing while they were increased for investors. Investment by landlords was seen as one causal factor in generating rapid house prices rises and as a consequence the right to claim mortgage interest as a tax relief was abolished in 1998. However, it was subsequently re-introduced as it was seen to have reduced the supply of rented accommodation. On the supply side, a number of measures were taken which sought to facilitate development in a speedier and more efficient manner. In 1998 the Serviced Land Initiative was introduced which funded the fast tracking of serviced land provision (MacCabe, 2003). With regard to planning, attempts were made to make planning more efficient, more planners were employed and policies on increased residential densities were introduced. However, one of the key responses by government has been to introduce a number of schemes that provide subsidised affordable housing for purchase, mainly to first-time buyers.

Subsidised Affordable Housing

In response to the second Bacon report, in which Bacon singled out affordable housing as one of the major problems in the Irish housing market, the government introduced an Affordable Housing Scheme in 1999. Since 1999 there have been two additional affordable housing schemes, one being Part V of the Planning and Development Act which again seeks to provide subsidised private housing and the other emanating from the latest national wage agreement, *Sustaining Progress*, where there is a commitment to develop 10,000 affordable houses on land owned by the central as opposed to local state (Government of Ireland, 2003). The 1999 scheme is intended to provide for the building of new houses in areas where house prices have created an affordability gap for first time buyers. Under the scheme local authorities will build houses, on land available to them, which will be sold at cost price to eligible candidates. In effect this is done by subsidising the land element of the house price. Where a local authority purchases land for this scheme, central government reimburse the local authority with a site subsidy. Where a local authority uses land that it already owns then a reduced subsidy applies, depending on how long the land has been owned. To be eligible for the affordable housing scheme, households must be covered by one of the following categories:

- Those in need of housing and whose income satisfies the income tests below
- A person whose application for local authority housing has been approved by the local authority
- A local authority tenant or tenant purchaser who wishes to buy a private house or to return their present house to the local authority, or
- A tenant for more than one year of a house provided by a voluntary body under the Rental Subsidy Scheme who wishes to buy a private house and return their present house to the voluntary body.

An income test applies only to the first category, and households qualifying under this category must be first time buyers, must be in permanent employment for at least one year prior to application, and their income for the previous tax year must be within the following limits as of July 2004. For a single income household, the gross income (before tax) in the last income tax year must be €36,800 or less to qualify. For a two income household to qualify, the gross income (before tax) of the higher earner in the last income tax year, multiplied by 2.5, plus the gross income of the other earner in the last tax year, must sum up to €92,000 or less to qualify. Houses are purchased under the affordable housing scheme with a loan provided by the local authority of up to 97 per cent of the sale price subject to a minimum deposit of 3 per cent and a maximum mortgage of €165,000. Loans are advanced over a 30-year term, with the option of a variable or fixed (for the first 5 years) interest rate, and cannot exceed 97 per cent of the sale price. The amount of the loan to be provided in individual cases is determined by the local authority having regard to household circumstances and the capacity of the household to meet repayments on the loan, the latter being subject to the repayments being no more than 35 per cent of the households' net income.

If the applicant cannot afford the monthly repayments on an annuity loan, then they can enter into a shared ownership agreement with their local authority. The applicant will initially acquire a minimum 40 per cent share and not more than 75 per cent share in an affordable house and rent the remainder from the local authority, with an undertaking to acquire the remaining equity within a 30-year period. The local authority will determine what share of the house the applicant will buy out initially based on the household's monthly repayments not exceeding 35 per cent of their net income. If the owners of a house purchased under the scheme wish to sell their house before the expiration of 20 years from the date of purchase, the person selling the property must pay the local authority a percentage of the proceeds of the sale. This is expressed as the percentage difference between the sale price and the market value of the house. This amount will be reduced by 10 per cent each year after you have owned your home for ten years. The household is free to sell their house after 20 years without having to pay anything to the local authority.

As Table 13.3 shows, since the introduction of the 1999 affordable housing scheme to the end of December 2003, a total of 2,804 affordable houses were completed nationally (Department of Environment, Heritage and Local Government, 1995-2003). The provision of affordable housing units under the scheme was slow to begin with, with only 272 units completed in 2001, and 882 by the end of 2002. Provision sped up in 2003, and a further 1,200 units were in progress at the end of the year. However, the rate of completion has been slow and by no way close to meeting the demand as determined by eligibility criteria (Downey, 2003).

Ten local authorities have no completions to date, and fourteen have no future schemes. Fingal County Council, in the Dublin area, completed by far the largest number of affordable houses, with 633 completions to date. Almost 86 per cent of the households that purchased an affordable house in 2003 qualified for the scheme via the income criteria, suggesting significant demand from those in low income employment, with little demand from those in local authority rented housing. However, as each local authority has a distinct allocation scheme, the figures may be influenced by different allocation practices. Almost half of the successful applicants in the scheme earned in excess of €25,000. Almost half of the affordable housing units sold in 2003 were in excess of €155,000, although the statistics do not give the upper range of prices. Nonetheless, in comparison to new house prices on the open market, the price of these affordable houses is significantly less. Ongoing research by the authors on the 1999 scheme suggests that there are significant differences in demand for this scheme. For example, in the Dublin City Council area, there has been high demand for recent schemes, with substantial surpluses of applicants for particular schemes. However, in Waterford City Council, the demand for affordable housing via this scheme is quite low, with the local authority finding difficulty in selling a number of small schemes. While there are many factors that account for this, the price differential

Table 13.3 Affordable house completions: State

	No. of houses provided	No. of houses in progress at end of year	Houses proposed at end of year
1999	40	96	784
2000	86	381	2,277
2001	272	1,177	4,168
2002	882	1,907	2,524
2003	1,524	1,209	1,481
Total	2,804		

Source: Department of the Environment, Heritage and Local Government (1995-2003)

between open market and affordable scheme is one such factor. In Waterford, the differential was of the order of only 3 per cent, while being significantly higher in the Dublin area.

The second affordable housing scheme stems from Part V of the Planning and Development Act 2000, which requires that local authorities produce Housing Strategies which plan for the provision of social and affordable housing. Under Part V each local authority must adopt a housing strategy that in turn must reserve a percentage (not more than 20 per cent) of the land zoned for residential development in the development plan to meet social and affordable housing needs. The division of this 20 per cent is determined in the housing strategy and has varied considerably across local authorities, with Dublin City for example specifying that there be a 50:50 split, but many other local authorities specifying higher percentages for affordable housing. For each individual planning permission, up to 20 per cent of either the land, serviced sites or completed dwellings must be transferred to the local authority for social and affordable housing. The key economic mechanism here is that such land must be transferred to the local authority at its existing use value rather than market value, thus making it possible for the local authority to provide subsidised affordable housing and also to construct social rented housing with a minimum site cost. A recent amendment to the Act in 2002 now allows the developer to transfer land, houses or sites at alternative locations to the site on the planning application or to make a payment in lieu of the land cost to the local authority. Because of lead-in times and the inherent complexity of the scheme at planning application stage, to date there has not been much progress in developments under Part V, with a total of 88 affordable units and 75 social rented units being provided so far. The third affordable scheme emanates from the most recent national partnership agreement, *Sustaining Progress*, which proposes that 10,000 affordable housing units be provided on state-owned land as opposed to local authority owned land (Government of Ireland, 2004). So far, no units have been provided under this scheme. Overall, the progress on the three schemes has been relatively slow and it is difficult to measure the impact of the schemes relative to demand for such subsidised affordable housing. However, the evidence that does exist shows that the affordability issue has not gone away, with house prices continuing to rise, seemingly inexorably, with incomes and general prices rising at much lower levels. Certainly in the Dublin region, the evidence suggests that the demand for such housing seems to be very high.

Conclusion

For most of the past decade the issue of affordability has been a vexed and controversial one. This chapter has sought to show that both conceptually and technically, affordability is difficult to measure. Moreover, it has sought to demonstrate that the available data, which seeks to measure affordability, is quite limited. Nonetheless, the evidence does suggest that there are affordability problems, especially in the Dublin and Eastern regions and most likely in other large urban areas. The policy responses to date have had limited impacts. While supply is at historically high levels, there is some evidence to suggest that a significant proportion of this supply has been for the second home market. If this is the case, then it is a form of market failure. Supply in the Dublin region, where affordability problems are at their worst, has only recently increased significantly. More specifically, the three affordable housing schemes have had only modest impacts to date, although recent research suggests high levels of demand for such subsidised housing in the Dublin area, given that anecdotal evidence would suggest that the key reason many commuters choose to live at such distances from the city is driven by the lack of housing affordability near the core.

One of the more interesting aspects of all three schemes is that in each of them the land or site costs is subsidised in some manner. In the 1999 scheme, central government pay local authorities a site subsidy; in the Part V scheme developers effectively provide a site subsidy; and in the most recent scheme central government will provide land they own for the development of affordable housing. Many analysts, including the Central Bank, have argued that between 40 per cent and 50 per cent of the sale price of a new house in Dublin is comprised of the land cost (Central Bank of Ireland, 2003). This may in part explain the significant geographical differences in new house prices as well as explaining in part the high cost of new housing. Earlier in 2004 an all-party committee of the Irish Parliament investigated this core issue and recommended that the government introduce some form of land betterment in order to control land prices and to gain some of the betterment for the community (Government of Ireland, 2004). While this issue of land values may indeed hold the key to house prices, given the vested interests involved in the development of high cost housing, it remains to be seen whether the government will in fact take any radical action of this type. Until that is done, the affordability problem remains and, in any future housing boom, is likely to return and further thwart government proposals to consolidate urban areas, and in particular, Dublin.

References

Bacon, P. and Associates (1998) *An Economic Assessment of Recent House Price Developments*, Dublin: Stationery Office.

Bacon, P. and Associates (1999) *The Housing Market: An Economic Review and Assessment*, Dublin: Stationery Office.

Bacon, P. and Associates (2000) *The Housing Market in Ireland: An Economic Evaluation of Trends and Prospects*, Dublin: Stationery Office.

Central Bank of Ireland (2003) *Quarterly Bulletins*, Dublin: Central Bank of Ireland.

Central Statistics Office (1996) *Census 1991: Vol 10 – Housing*, Dublin: Stationery Office.

Central Statistics Office (2004a) *Census 2002: Vol 10 – Housing*, Dublin: Stationery Office.

Central Statistics Office (2004b) *Quarterly National Household Survey: Housing and Households*, Dublin: Stationery Office.

Department of the Environment and Local Government (1998) *Action on House Prices*, Dublin: Department of the Environment and Local Government.

Department of the Environment and Local Government (1999) *Action on the Housing Market*, Dublin: Department of the Environment and Local Government.

Department of the Environment and Local Government (2000a) *Action on Housing*, Dublin: Department of the Environment and Local Government.

Department of the Environment and Local Government (2000b) *Part V of the Planning and Development Act 2000, Guidelines for Planning Authorities – Housing Supply*, Dublin: Stationery Office.

Department of the Environment and Local Government (2000c) *Part V of the Planning and Development Act 2000, Housing Supply – A Model Housing Strategy and Step-By-Step Guide*, Dublin: Stationery Office.

Department of the Environment, Heritage and Local Government (1995-2003) *Annual Housing Statistics Bulletin*, Dublin: Stationery Office.

Department of the Environment, Heritage and Local Government (2004) *Housing Policy*, http:// www.environ.ie.

Department of Social and Family Affairs (2003) *Statistical Information on Social Welfare Services*, Dublin: Stationery Office.

Downey, D. (2003) 'Affordability and Access to Irish Housing: Trends, Policy and Prospects', *Journal of Irish Urban Studies*, Vol. 2(1), pp. 1-24.

Drudy, P.J. and Punch, M. (2001) 'Housing and inequality in Ireland', in Cantillon, S., Corrigan, C. and Kirby, P. (eds) *Rich and Poor: Perspectives on Tackling Inequality in Ireland,* Dublin: Combat Poverty Agency and Oak Tree Press.

Drudy, P. and Punch, M. (2002) 'Housing Models and Inequality: Perspectives on recent Irish Experience', *Housing Studies*, Vol. 17(4), pp. 657-672.

Duffy, D. (2004) 'A Note on Measuring the Affordability of Homeownership', *Quarterly Economic Commentary*, Dublin: Economic and Social Research Institute, pp. 71-78.

European Union (2002) *Housing Statistics in the European Union, 2001*, Finland: Ministry of the Environment.

Fahey, T. (2004) 'Housing Affordability? Is the Real Problem in the Private Rented Sector?', *Quarterly Economic Commentary*, Dublin: Economic and Social Research Institute, pp. 79-96.

Fahey, T., Nolan, B. and Maitre, B. (2004) *Housing, Poverty and Wealth in Ireland*, Dublin: Institute of Public Administration and Combat Poverty Agency.

Focus Ireland, Simon Communities of Ireland, Society of Vincent de Paul and Threshold (2002) *Housing Access for All? An Analysis of Housing Strategies and Homeless Action Plans*, Dublin: Focus Ireland.

Government of Ireland (2000a) *Planning and Development Act, 2000*, Dublin: Stationery Office.

Government of Ireland (2000b) *Ireland: National Development Plan: 2000-2006*, Dublin: Stationery Office.

Government of Ireland (2003) *Sustaining Progress: Social Partnership Agreement 2003-2005*, Dublin: Stationery Office.

Government of Ireland (2004) *The All-Party Oireachtas Committee On The Constitution Ninth Progress Report: Private Property*, Dublin: Stationery Office.

Hancock, K.E. (1993) 'Can Pay? Won't Pay? Or Economic Principles of "Affordability"', *Urban Studies*, Vol, 30(1), pp. 127-145.

MacCabe, F. (2003) 'Supply constraints and serviced land development supply in the Dublin region: A Review of the Projections and Recommendations of Bacon III', *Journal of Irish Urban Studies*, Vol. 2(1), pp 55-64.

McCarthy, C., Hughes, A. and Woelger, E. (2003) *Where Have All The Houses Gone?* Research Report, Dublin: Davy Stockbrokers.

McNulty, P. (2003) 'The Emergence of the Housing Affordability Gap', *Journal of Irish Urban Studies*, Vol. 2(1), pp. 83-90.

Memery, C. (2001) 'The Housing System and The Celtic Tiger', *European Journal of Housing Policy*, Vol. 1(1), pp. 79-104.

National Economic and Social Council (1988) *A Review of Housing Policy*, Dublin: National Economic and Social Council.

Nolan, B., O' Connell, P. and Whelan, T. (eds) (2000) *Boom to Bust: The Irish Experience of Growth and Inequality*, Dublin: Institute of Public Administration.

O'Sullivan, T. and Gibb, K. (eds) (2003) *Housing Economics and Public Policy*, Oxford: Blackwell Science Ltd.

Watson, D. and Williams, J. (2003), *Report on the Irish National House Condition Survey 2000-2001*, Dublin: Economic and Social Research Institute.

PART III
CONCLUSION

Chapter 14

Towards a Sustainable Future for Irish Towns and Cities

Niamh Moore and Mark Scott

Introduction

> Sustainable development is not an abstract concept. It is about securing a better quality of life for all, both now and for future generations. It is about achieving a successful, stable economy, while creating a strong and inclusive society and protecting the environment. This is our challenge for the 21st century. It means understanding the impact of the decisions we make as governments, as businesses and as individuals, and in some cases, making difficult choices to get the balance right (DEFRA, 2003).

As countries have become more urbanised, the search for the optimal quality of life has become more important. Since the 1960s, the idea that this concept could in some way be measured and modelled has held much resonance with social scientists. Some theorists have defined measurable attributes including income and wealth, the state of the environment, health and education, a feeling of social order and social belonging and the availability of recreation as central to understanding the relatively abstract (Smith, 1975). However in more recent years, given the increased debate surrounding quality of life and its application to the sustainability agenda, Senecal (2002) has made an attempt to synthesise these measures into two key ideas. He argues in a Canadian context, but one that is applicable to most highly-developed countries that, fundamentally, quality of life is about the *living environment* and the patterns of inequitable advantages and opportunities. Secondly, drawing on previous work by Perloff (1969) he argues that in any debate, a key concern is the *natural environment of urban spaces*.

In developing a methodology that could be used by planners to effectively understand and thus improve quality of life issues, Myers (1988) argues for the adoption of a community-trend method, which would link the concept with the context within which it is both perceived and changed over time. He argues that quality of life is often used to encourage and promote economic development or growth, and is in turn altered by it. Given the profound economic, social and spatial transformations in Ireland over the last decade, described by *The Economist* (2004a) as a 'rags-to-riches story', quality of life issues in recent months and years have again begun to permeate both popular and political discourse. Although ranked in a more recent issue of *The Economist* (2004b) as the country with the highest standard of living in the world, this purely economic definition does not accurately reflect the perceived reality of life in Ireland today. It is therefore important to ensure the subtle difference between standard of living and quality of life is made clear. While the measures adopted by this publication include some of the attributes identified by Smith (1975) above, such as income and wealth, quality of life is a much broader concept and would include issues such as social cohesion and the quality of the built environment. Directed by Wellington City Council, a quality of life project undertaken by New Zealand's local authorities has identified a set of indicators which may be useful in an Irish context in attempting to identify how a changing socio-economic, demographic and spatial context has impacted on quality of life in this country. Table 14.1 identifies some of the key indicators now being used in New Zealand to comparatively identify areas for improvement and emulation, and it provides a brief description of how Ireland scores on each one. Although scoring well on some indicators such as economic development and civil and political rights, it becomes obvious that there are a number of areas that need increased or at least re-focused attention. This book has attempted to address a number of these key indicators, namely economic development, housing, the natural and built environment and social connectedness, on which the following general conclusions are based.

Sustainable Patterns of Development?

As the preceding chapters have illustrated, the island of Ireland provides a textbook case study of the manner in which socio-economic shifts result in the emergence of entirely new spatial forms. In the last decade, former agricultural land has been rezoned and urbanised with row upon row of new suburban estates while urban centres, formerly worthless, derelict, empty cores, have undergone a dramatic change in terms of both character and extent. In Chapter 7, Chris Paris has described the unprecedented growth experienced in the city of Derry (Northern Ireland) close to the border with the Irish Republic, since the resumption of political stability and the effect that this has had on neighbouring county Donegal (Republic of Ireland), which is now servicing the housing needs of this growing city. This theme was also considered in Chapter 8, which examined spatial relationships within the Greater Dublin Area, given that Dublin has mushroomed into the surrounding counties of Meath, Kildare and Wicklow. In this chapter Gkartzios and Scott identify the 'rural idyll with accessibility to the city' as the most obvious spatial settlement pattern emerging. They argue that the current pattern of development, which to some extent can be classified as an urban-rural migration within the GDA, appears to be constrained and controlled by the lines of road and rail transport. Whether this migration is inevitable or a result of residents disillusionment with the kinds of urban environment that

Table 14.1 Indicators of quality of life applied to Ireland

Indicator	*Ireland's record*
People	Republic of Ireland population of 3.95 million, 67.5% between 15-64 years; Northern Ireland population of 1.7 million, 61% of population between 15-64 years; In general, young urban population, although depopulation is becoming a problem in some parts of country.
Knowledge and skills	In the Republic of Ireland, education accounts for 14% of public expenditure; In Northern Ireland, education accounts for 22% of public expenditure; Levels of educational attainment are generally high.
Economic standard of living	In the Republic of Ireland per capita GDP is 10% above that of the four big European economies, but society is increasingly characterised by an increasing income gap; In Northern Ireland, standard of living was adversely affected by the 'Troubles' but has in recent years improved dramatically.
Economic development	Republic of Ireland average growth rate of 8% in 1995-2002, levelling at 4% in 2003, Northern Ireland average annual growth rate is 3%; this is 30% in excess of the EU15 average.
Housing	In both parts of the island, dramatic house price increases driven by economic development and political stability; property speculation and affordability are critical issues.
Natural environment	New policies attempting to balance economic growth and environmental issues.
Built environment	Under extreme pressure from population, infrastructure needs and property speculation.
Social connectedness	Under debate, with many believing that community and family values have significantly weakened in recent years, and social, economic and cultural marginalisation has increased.
Civil and Political rights	Republic of Ireland is a democratic republic, with the judicial system guaranteeing rights; Northern Ireland is a constitutional monarchy, civil and political rights issues have been redressed through recent political and legislative change.

already exist is perhaps a question for future research, but in Chapter 3 Craig Bullock argues, based on new empirical evidence, that any new model of urbanisation being considered in Ireland must include better planned, diverse and more attractive green space.

This is not simply an academic, but also a pragmatic prescription given that a large percentage of urban dwellers do not appear to use existing urban green space, due to its blandness or perceived lack of safety. In the context of the debate on brownfield regeneration, introduced by Moore, this is an important consideration, as recent guidelines issued by city planners dictate that all new developments in Dublin must include a provision of 10% green space. Rather than perceiving of brownfields as problem areas, local authorities need to consider the potential that these sites offer for radically altering the existing unsustainable patterns of development, that have resulted in enormous environmental, social and economic problems. This is reinforced by McEldowney et al., in their chapter on the Belfast Metropolitan Area, given the emphasis in the Regional Development Strategy for Northern Ireland on promoting brownfield over Greenfield, or edge-of-city, development. Building on the work of Breheny (1992), they argue that any debate on the accommodation of development must consider the links between urban form, transport and quality of life. This theme is also reflected in MacLaran's discussion of 'Edge city' formation in Dublin and the enormous problems that this poses from a long-term sustainability perspective. Given the multiplicity of nodes of employment (edges) and new residential areas, the existing transport system cannot cope and the Greater Dublin Area, like the Belfast Metropolitan Area, has become characterised as an unsustainable, sprawling and car-dependent urban region in part due to inappropriate and ineffective urban strategies.

The Policy Environment

A central focus of many chapters has been the problems that current policy has created and the attempts to rectify previous mistakes. As we suggested in the Introductory chapter, it has become clear that policy implementation is a key stumbling block to sustainability in Ireland, as the gap between government rhetoric and the situation on the ground continues to widen. Given the range of topics that have been considered, the difficulties appear to fall into two key areas. Firstly, there is little or no policy integration across levels of government and across central government departments, leading to policy replication and indeed in some cases policy contradictions. This was most effectively demonstrated in Chapter 8 in relation to counter-urbanisation in the Greater Dublin Area. Gkartzios and Scott have illustrated that while the National Spatial Strategy argues for residential development in the rural hinterland of Dublin being confined to identified towns and service villages, rural housing guidelines suggest a more permissive approach to dispersed housing in the countryside, including areas in close proximity to urban centres. Clinch has also demonstrated in Chapter 6 the need for co-ordination and integration in the realm of energy efficiency given that it is usually the concern of a number of government departments, and in an Irish context he suggests that an agency like Sustainable Energy Ireland may have a key coordinating role to play in terms of policy development and enforcement.

This leads on to the second difficulty, which is an apparent reticence to actually engage in policy enforcement, whether due to lack of resources of sheer lack of political will. In Chapter 2, Moore exemplifies the difficulties faced by local authorities in promoting brownfield regeneration given the weak nature of the legislation, the Derelict Sites Act that is supposed to provide a framework for action. Ellis et al. have remarked in general in relation to the embeddedness of LA21 within local authority structures that a similar problem arises. Because local sustainable development and LA21 are not statutory, they have a weakness and have therefore barely rated in terms of competition for and allocation of resources. The political will needed to address difficult issues and alter existing management and resource allocation structures is simply not evident. Nowhere has this been more evident that in the debate over affordable housing that has taken place in the Republic of Ireland over the last number of years, where a Ministerial amendment rendered the legislation almost totally ineffective. The end result according to Redmond and Kernan has been a rolling back of the opportunity to broaden access to housing, a situation that is likely to deepen over in coming years, and one that may result in increased social exclusion and a reduced sense of urban citizenship.

Citizenship Issues

> It is interaction, not place, that is the essence of the city and of city life (Webber, 1964, p. 147).

One of the key themes running through each chapter has been the role and engagement of different stakeholders in the development and planning process. The importance of fitting the objectives of policy to the needs of the urban dweller or user was highlighted in Chapter 5, where McEldowney et al. argue for active consultation in the development of strategic metropolitan plans. This entreaty is based on the key requirement of local sustainable development and LA21 to engage all stakeholders to ensure legitimacy of both the process and outcome, and indeed resolve the kinds of tensions that currently exist. This bottom-up approach is critical given the apparent gap that has emerged between authorities and urban dwellers in terms of the preferred future city. This is developed by Ellis et al. in Chapter 9 where they highlight the need for dialogue to occur between local authorities and citizens, for links between governance and sustainability to be explored and they conclude that given their findings, they are indeed hopeful that active citizenship is emerging. In his discussion of the URBAN II Case study in

North Belfast, Murtagh supports the benefits of a partnership approach, which is critically analysed by Russell et al. in their discussion of the effectiveness and activity of residents associations, as examples of active citizenship. Nonetheless, in her discussion of participatory democracy in North Belfast and Ballymun (Dublin), Jenny Muir questions whether in the current political climate, partnership is being promoted for the right reasons or whether it is in fact part of a much larger hegemonic project of legitimisation. In this regard, Murtagh's discussion on the 'dark side of difference' and Russell et al.'s analysis of the formation of social capital as a means to exclude otherness, provides a useful critique of contemporary citizenship debates. These are timely observations given the increasing debate surrounding broader citizenship issues in Ireland, which has sparked a contentious debate on who belongs and who does not? Drawing on the ideas of Pahl (1974), the key question for all stakeholders in the current economic climate should be whose city is being created? Recent criticism of the effectiveness of existing local partnership structured and indeed RAPID, a central government programme for urban revitalisation in the most disadvantaged areas, has called into question the extent to which current 'partnership' arrangements are actually fulfilling their objectives, and achieving sustainable long-term objectives.

Conclusion: Achieving a Better Quality of Life – Lessons from Ireland

As we have reiterated at many points throughout this discussion, Ireland's urban areas have experienced an unprecedented and perhaps unparalleled transformation over the last decade, both by rapid economic growth in the case of the Republic of Ireland, and through political processes and change in Northern Ireland. This book has attempted to illustrate the environmental consequences of new patterns of urban development, while also identifying new forms of exclusion that have emerged in the contemporary Irish city. Recent years have witnessed considerable progress in developing new spatial policy frameworks for both parts of the island, in particular the National Spatial Strategy (NSS) for the Republic of Ireland and the Regional Development Strategy (RDS) for Northern Ireland – both of which emphasise sustainable development as the primary policy goal. However, policy implementation and operationalising the concept of sustainability beyond rhetoric has proved much more elusive and selective. Similarly, new approaches to urban governance have become commonplace with urban management initiatives advanced in 'partnership' with local communities. But key questions remain in relation to access to partnership and participatory processes, and the varying capacities of urban communities to influence policy outcomes. In this analysis, a number of key lessons can be developed from the Irish urban experiences that have relevance beyond the island of Ireland, not least in other societies undergoing rapid economic and political transformations.

Firstly, developing effective mechanisms for policy implementation remains an enduring quest. Traditional regulatory approaches to managing urban spatial change have been less than successful in securing more sustainable patterns of urban development. Although policy prescription increasingly favours a sustainable urban development agenda the emerging geography of Ireland displays distinctly unsustainable trends, suggesting a gap between policy intentions and actual outcomes. In this context, there may be considerable potential for both parts of the island to learn from practice in the other jurisdiction. For example, in the Republic of Ireland, there is currently no effective mechanism for ensuring that national and regional spatial goals are translated into statutory local authority development plans – a significant gap in policy implementation. However, in contrast in Northern Ireland, local area plans are scrutinised and given a 'certificate of conformity' by the Department of Regional Development to ensure consistency with the Regional Development Strategy. A further area of potential transfer relates to the use of market-based instruments (primarily tax incentives) as a tool for urban renewal, often favoured in the Republic of Ireland. This approach may offer potential for encouraging inner city residential development and increasing residential densities in Northern Ireland, as identified by McEldowney et al. in Chapter 5. Furthermore, a clear need exists to develop evidence-based approaches to policy-making to address the gap between policy intentions and outcomes. For example, although policy-makers increasingly favour introducing higher residential densities in urban areas, evidence from Belfast and the Greater Dublin Area suggests that consumers continue to express a preference for lower density residential areas. In this regard, research to examine consumer residential decision-making would allow policy-makers to develop appropriate policy responses.

Secondly, addressing the 'disconnect' between environmental and spatial issues with social and economic objectives presents a considerable challenge. Chapter 1 stressed the importance of sustainable development as a multi-dimensional and integrative concept involving the pursuit and reconciliation of economic, social and environmental objectives. However, translating the concept of integrated sustainable development into implementation has been elusive and contested at the point of delivery. This suggests that, as Whitehead (2003) argues, research is needed to address the sustainable city as an object of political contestation and struggle – for example, both McEldowney et al. (Chapter 5) and Russell et al. (Chapter 12) outline negative outcomes of community participation and the clear influence of NIMBY attitudes in opposing local (physical) development, particularly high residential density schemes and development proposals perceived as a threat to existing property values. As already highlighted, there is a clear need for integrated approaches to policy-making from national to local level. For example, at a local level land-use development plans in both Northern Ireland and the Republic of Ireland are developed largely in isolation of local development initiatives and urban partnerships, suggesting the need for new arenas to formulate integrative policy discourses. In the Republic of Ireland, policy integration is hampered by the absence of an explicit national urban policy, which may provide a common frame of reference for various policies and programmes impacting on urban areas.

Thirdly, and related to both policy implementation and integrative policy-making, lies the challenge of reinvigorating multi-level governance, and the need to broker connections within and between cities and between urban and rural areas. Within the Republic of Ireland, progress has been made in attempting to coordinate local government and local development partnerships with the introduction of County and City Development Boards in 2000. However, partnerships should not be viewed as synonymous with participation and the ability and capacity of the more informal networks of residents' and community groups requires further attention within research. Regional governance remains poorly developed in the Republic of Ireland, with limited responsibilities for the State's Regional Authorities. A recent opportunity has emerged following the publication of the National Spatial Strategy, whereby Regional Authorities have been involved in developing regional planning guidelines. In the case of the Greater Dublin Area, though, the administrative area (the Dublin and Mid-Eastern Regional Authorities) does not correspond to the emerging functional area of the city-region, which has extended far beyond the regional administrative boundaries. However, the scope for enhancing regional governance may be limited given an absence of regional identity in Ireland generally, and the intense localism that characterises Irish politics. In Northern Ireland, the future of regional governance is inextricably linked to ongoing political negotiations and the peace process. Devolution to the Northern Ireland Assembly has been a stop-start affair, however, at the time of writing, there has been renewed optimism that devolved political structures may again be workable. There is also uncertainty relating to the future of local government in Northern Ireland. Currently twenty-six district councils operate in the Province, which many commentators have argued is too many for a small region such as Northern Ireland, and there is much speculation concerning the future scale and role of local government.

To date, inter-regional governance between Northern Ireland and the Republic of Ireland has been limited. However, as discussed by Paris (Chapter 7), since the demilitarisation of the border, housing and labour markets have increasingly operated within a cross-border framework. The case of Derry/Londonderry is particularly illustrative as the city has expanded not only beyond the local authority administrative boundary, but has also sprawled into County Donegal in the Republic of Ireland. Although the need to consider the City of Derry in its cross-border context is recognised in both the NSS and RDS, further attention is required to address measures for effective urban management of the city that cuts across traditional administrative boundaries, political jurisdictions and planning regimes. In recent years, the concept of enhancing the Dublin-Belfast corridor along the east of the island has commanded considerable interest within the policy community. Within a European context, the Dublin-Belfast corridor perhaps offers the potential to explore the notion of polycentric urban development on a cross-border and collaborative basis, providing the critical mass to improve the economic competitiveness of each region (for example, access to labour markets). However, a coherent strategy for cross-border spatial policy has yet to emerge. As Murray argues, the island of Ireland perspective is a contested geopolitical construct, and while both spatial strategies goes a certain distance, they stop short of embracing the territorial logic of comprehensive all-island spatial planning, although perhaps this is not what is required:

> At a practical level the way ahead arguably lies with working-up a shared action-oriented agenda for sustainable development ... This phrase, more than any other, watermarks the pages of the RDS and NSS, not least in regard to future patterns of settlement development. Its identification and selection as 'the key strategic choice' could well create the foundation for the next phase of policy succession on the island of Ireland (Murray, 2004, p. 240).

This echoes Ellis et al.'s arguments (Chapter 9), which call for an island ethic of sustainability based on citizenship and cross-border collaboration. It is hoped that this book, with its cross-border and inter-disciplinary perspective, can contribute to this policy debate.

References

DEFRA (2003) *Achieving a better quality of life: Review of progress towards sustainable development: Government Annual Report 2003*, London: Stationery Office.

Lynch, K. (1960) *The Image of the City*, Cambridge, MA: MIT Press.

Murray, M. (2004) 'Strategic Spatial Planning on the Island of Ireland: Towards a New Territorial Logic?', *Innovation*, Vol. 17, pp. 227-242.

Myers, D. (1988) 'Building knowledge about quality of life for urban planning', *Journal of the American Planning Association*, Vol. 54, pp. 347-358.

Pahl, R. (1975) *Whose city? And further essays on urban society*, Harmondsworth: Penguin.

Perloff, H.S. (1969) *The Quality of the Urban Environment*, Baltimore: Johns Hopkins University Press.

Senecal, G. (2002) 'Urban spaces and quality of life: Moving beyond normative approaches', *Horizons*: *Policy Research Initiative*, Vol. 5(1), pp. 20-22.

Smith, D.M. (1975) *Patterns in human geography: An introduction to numerical methods*, Devon; David & Charles Holdings.

Webber, M.M. (1964) *Explorations into urban structure*, Philadelphia: University of Philadelphia Press.

Index

Abercrombie, Patrick, 52
Agenda 21 (UN), 199
  See also Local Agenda 21
air quality, 97–99, 100–101
Amsterdam, Treaty of (1997), 30
Area Committees, 12
assimilation of ethnic groups, 180–81
Athlone as regional gateway, 16

Bacon reports on housing, 149, 150, 244
Barrington Committee on local government (1991), 11, 218
Belfast city
  consumer mobility survey, 184–85
  decentralisation, 79, 86, 119
  development of, 4–6
  Housing Strategy, North Belfast, 202–10
  industrial manufacturing, 4
  population, 4, 5, 79, 118
  regional gateway, 16, 79
  segregation, ethno-religious, 5, 182–87
  Titanic Quarter development, 88
  URBAN II Community Initiative Programme (2000-2006), 187–94
  See also Belfast Metropolitan Area
Belfast Metropolitan Area (BMA)
  Belfast Metropolitan Area Plan (BMAP), 75, 80
  Belfast Metropolitan Transport Plan (BMTP), 82
  car travel, 78, 84–87
  housing growth, 79–80
  population, 4, 78
  public consultation process, 75, 87–89
  Strategic Employment Locations, 86
  transport policy and land use, 75–91
  travel behaviour survey, 84–87
  See also Belfast city
'Better Local Government, A Programme for Change' White Paper (1996), 12–13, 164
border counties
  growth between Derry and Co. Donegal, 114–29
brownfields, regeneration of, 29–47, 78, 80
Bruntland Report on sustainable development, 199
'bungalow blitz'. *See* housing development, dispersed in rural areas

Canada
  brownfields, regeneration of, 34–35, 46
car travel, 71–73, 75–91, 99–102, 151–52
'Celtic Tiger'. *See* Ireland, economic development
Charter of European Towns and Cities Towards Sustainability, 1
citizenship. *See* 'civil society' participation; residents' associations
'civil society' participation, 10, 13–14, 171–72, 180, 197–210, 213–31
CLARINET Working Group, 30
collective action. *See* residents' associations
community interests. *See* 'civil society' participation; local government, partnerships; residents' associations
commuting, car-based. *See* car travel
contamination of land
  liability for, 39–40
  remediation of, 35, 42–44
  See also brownfields, regeneration of
Cork city
  population, 2, 17, 119
  regional gateway, 16
Corporate Policy Groups, 12
corridors, transport. *See* regional planning
counterurbanisation, 4, 76, 118, 132–53
County/City Development Boards, 12–13, 164
Coyne, Peter, 42

cultural capital, 213–31
Custom House Docks development, Dublin, 41, 44, 193
Czech Republic
brownfields, regeneration of, 31–34

decentralisation
cities, 63–64, 79, 86, 118–19, 132–37
Civil Service, 17
industrial, 64
office stock, 67–69
retailing, 64
workforce, 70
Derelict Sites Act and Register, 38–39
Derry city
expansion into Co. Donegal, 114–29
population, 115–17, 127
regional gateway, 16, 20, 79
segregation of housing, 115
developers
fiscal incentives for property development, 47, 66, 67, 70
investment demand, 70
Donegal, Co.
cross-border growth with Derry city, 114–29
population change, 126
Draft Development Plan (Dublin City Council), 37
Dublin city
Ballymun Regeneration programme, 202–10
Custom House Docks development, 41, 44, 193
decentralisation, 63–64, 119
density, 3
Derelict Sites Act and Register, 38–39
Grand Canal Dock 'brownfield' development, 40–46
office development, 3, 64–70
population, 1–2, 62, 118
retail parks' development, 3, 64
'road-pricing charge' proposal, 108
suburbs, growth of, 3, 50, 60–73
sustainability of, 3, 70–73
See also Greater Dublin Area
Dublin City Council, 37, 39, 202
Dublin Corporation, 39, 62–63
See also Dublin City Council
Dublin Docklands Development Authority, 39, 40–46
Dundalk
population, 17
regional gateway, 16

Eastern Europe
environmental regeneration, 31
mining industry, 32–34
economic development of Ireland, 2, 7, 94, 203
'edge city' developments, 3, 61, 62, 70–73, 101
energy efficiency policy measures
building sector, 104–6
residential sector, 104–6, 108–10
transport sector, 102, 106–8
energy use, 94–111
consumption by sector, 94, 95
consumption by transport sector, 100
consumption in residential sector, 103
environmental implications, 94–99
Engineering and Physical Sciences Research Council Sustainable Cities project. *See* European Sustainable Cities Project
Environment and Development, UN Conference on. *See* Rio Summit 1992
Environment and Development, World Commission on, 6, 199
environmental impacts, 94–99
Environmental Protection Agency, 38, 42–43
environmental protocols, targets of, 96–97
European Spatial Development Perspective (1999), 15
European Sustainable Cities Project (1996), 29, 75, 84–87
European Union (EU)
'compact city' concept, 20, 77
counterurbanisation trends, 134–36
economic growth among Member States, 7, 8, 94
Emissions Trading Scheme, 109
Energy Performance Directive, 108
industrial land regeneration (brownfields), 30
local partnership programmes, 11, 13, 182

spatial development policy, 15, 30, 77
sustainable development policy, 9–10, 30
urban – rural migration, 134–36
urban development policy, 9–10, 77
Water Framework Directive, 46

Finland
urban – rural migration, 135
fiscal incentives for property development, 47, 66, 67, 70
fuel poverty, 104, 110

Galway city
population, 2
regional gateway, 16, 17
Garden City Movement, 50
'garden suburbs', development of, 50, 62
gateways, regional. *See* regional planning
Germany
brownfields, regeneration of, 30
Gothenburg Protocol, targets of, 96, 97, 98
Grand Canal Dock 'brownfield' development, Dublin, 40–46
Greater Dublin Area (GDA)
economic growth, 2, 7–8
expansion of, 3–4, 60–73
motorway development, 3, 71, 101
population growth, 2, 7, 62
Regional Planning Guidelines for GDA, 19, 37, 138
spatial strategy for, 16, 19, 138–39
See also Dublin city
green space, urban, 49–58
greenfield sites, development of, 30, 45, 46, 47, 66, 67, 78, 80, 86
greenhouse gases, 95–99, 101
GREENSPACE survey, 52, 53–58
greenways, 52

historic buildings, planning restrictions for, 69
housing development
affordable housing, 235–48
Bacon reports, 149, 150, 244
change in housing systems, 121–23
Derry – Donegal border zone, 115–17, 123–28
dispersed in rural areas, 138–41
Dublin suburbs' growth, 62–63
'edge city' developments, 101–2
government policy, 138–41, 235–36, 244–47
house prices, 148–50, 237–44
Leinster province, 145, 147–48
regeneration programmes (Belfast and Dublin), 202–10
social housing provision, 121–23
urban expansion of Belfast, 79–80
Howard, Ebenezer, 50
hubs, regional. *See* regional planning
Human Settlement, UN Conference on. *See* Rio Summit 1992

Industrial Development Authority, 66
industrial estates, development of, 63–64
industrial land, redevelopment of. *See* brownfields, regeneration of
Interdepartmental Task Force on the Integration of Local Government and Local Development Systems, 12
Ireland, island of. *See* Ireland, Republic of;
Northern Ireland
Ireland, Republic of
brownfields, regeneration of, 37–47
cities, population of, 1–2, 118–20
cross-border growth, 20, 114–29
economic development, 2, 7, 94, 203
housing development, 62–63, 101–2, 121–23
local government system, 10–14
population changes, 2–3, 7, 118–20, 141–45, 146
regional development/spatial strategy, 15, 18,
16–20, 37
sustainable development policies, 37, 163–67

Jacobs, Jane, 52
Kildare, population changes in county, 145, 146
Kyoto Protocol, targets of, 96–97, 109

labour force, movement of, 70
Laois, population changes in county, 145
LEADER programme (EU), 11, 13

Leinster, province of
house prices, 148–50
housing development, 101–2, 145, 147–48
population changes, 141–45, 146
Letterkenny as regional gateway, 16
Liebskind, Daniel, 44
Limerick city
population, 2, 17
regional gateway, 16
Local Agenda 21 (LA21), 160–75
local authorities
commercial rates for office development, 69, 71
competition between, 3
derelict sites and brownfields, 38–39
financial support (subvention) from central government, 3, 71
green space management by, 52
housing strategies, 243–47
institutional capacity for sustainable development, 170–71
sustainable development and, 160–75
See also local government
local government
Barrington Committee (1991), 11, 218
'Better Local Government' programme, 12–13, 164
'civil society' participation, 10, 13–14, 171–72, 197–210, 213–31
partnerships for local development, 11–12, 13, 14, 182, 198–99, 219
reform of, 12–13, 219
services provided, 11
traditional system, 10–11, 218
Londonderry. *See* Derry
Louth, population changes in county, 142, 144, 145

management of cities. *See* urban management
Meath, population changes in county, 144, 145
metropolitan de-concentration, 118–19
See also counterurbanisation
mining industry, 32–34
motorway development, 3, 71, 101
Mullingar as regional gateway, 16
National Anti Poverty Strategy, 194
National Development Plan (2000-2006), 16, 110, 114
National Hazardous Waste Management Plan, 38
National Land Use Database (UK), 36
National Spatial Strategy (2002-2020), 2, 15, 16–19, 37, 77, 138–41
Neighbourhood Renewal, Northern Ireland Strategy for (2003), 187, 192
Netherlands
brownfields, regeneration of, 30
urban – rural migration, 135
New Zealand
'quality of life' indicators, 254
Noble Index, 187
Northern Ireland
car culture, 78
cities, population of, 118–20
cross-border growth, 20, 114–29
economic development, 7
housing development, 79–80, 115–17, 121–28
local government partnerships, 13, 198
Neighbourhood Renewal Strategy, 187, 192
peace process, 5, 187
political factors, 4
population changes, 2–3, 7, 115–17, 118–20
public consultation process, 15, 87–89
regional development/spatial strategy, 15, 16, 17, 20, 77–78, 79–83
sustainable development policies, 162–63, 166
transport policy and land use, 75–91
urban development, 4–6
Urban Regeneration Strategy, 198
See also Belfast city; Belfast Metropolitan Area
Northern Ireland Act (1998), 187

office development in suburbs, 3, 61, 64–70
'one-off' housing. *See* housing development, dispersed in rural areas
open space. *See* green space, urban

Paris, vehicular movement in, 72
parks. *See* green space, urban
PEACE programme (EU), 198
Planning and Development Act (2000), 18, 219, 243
political pragmatism, 15–16, 17, 37
population. *See* Ireland, Republic of; Northern Ireland; *and individual places*
Poverty 3 programme (EU), 11
Property Surveys, IAVI Annual, 150
public consultation process, 15, 75, 87–89, 203–5
public participation
  in local governance, 12–14
  in urban regeneration, 197–210
  See also residents' associations
public surveys on
  consumer mobility (Belfast), 184–85
  environment and quality of life issues, 136–37
  green space, 53–58
  local sustainable development, 165
  travel behaviour (Belfast and Edinburgh), 84–87
public transport, use of, 71, 72, 106–7, 151

Regional Authorities, 18–19
Regional Development Strategy for Northern Ireland (2000-2025), 15, 17, 79, 83
regional planning, 15–19, 37, 79–83, 138–39
  cross-border potential, 20, 258–59
  gateways, 16–17, 18, 79
  hubs, 18, 79
  transport corridors, 79
Regional Planning Guidelines, 19, 37, 138
Residential Density: Guidelines for Planning Authorities (1999), 51, 77
Residents' associations, 213–31
retail parks, development of, 3, 64
Rio Summit 1992 (UN), 9, 37, 159, 160, 161, 199
Rio+10. *See* Sustainable Development, World Summit on
road improvements, 71, 106–7
Rural Housing, Draft Guidelines for Sustainable (2004), 140

segregation, social and ethnic, 179–94
Shannon as regional gateway, 16
shopping centres, development of, 64
Sligo
  population, 17
  regional gateway, 16
social capital, 213–31
social exclusion, 181–82
social housing. *See* housing development, social housing
spatial strategies, 14–20
  See also European Union; Ireland, Republic of; Northern Ireland
Strategic Policy Committees, 12
suburbs
  Dublin's growth of, 3, 50, 60–73
  higher density development, 51
  industrial estates in, 63–64
  low density model, 51, 63
  office development in, 3, 61, 64–70
  retail parks in, 3, 64
Superfund programme (USA), 35, 40
surveys. *See* public surveys
sustainable development
  cross-border potential, 172–75
  definition of, 6, 159, 167–68
  Dublin 'edge city' sustainability, 70–73
  EU policy, 9–10, 30
  future for Ireland of, 253–61
  implementation of policies (Belfast case study), 87–89
  local authorities and, 160–75
  public participation in, 197–210
  'quality of life' indicators, 251–53
  urban form and, 76–78
  urban management and, 6–10
Sustainable Development: A Strategy for Ireland (1997), 37, 163
Sustainable Development, World Summit on (Johannesburg, 2002), 162
Sustainable Energy Ireland, 108, 110
sustainable participation, 199–210

traffic congestion, 51, 71, 100, 102

transport corridors. *See* regional planning
transport habits. *See* car travel
transport planning, 106–7
transport policy and land use
  Northern Ireland case study, 75–91
travel behaviour surveys (Belfast and Edinburgh), 84–87
Tullamore as regional gateway, 16

United Kingdom (UK)
  brownfields, regeneration of, 35–36, 46
  Neighbourhood Renewal Areas, 182, 187
  sustainable development and local authorities, 163–64
  urban – rural migration, 135
  Urban Renaissance Report (1999), 77
United States of America (USA)
  brownfields, regeneration of, 35
  'civic environmentalism', 171
  Community Empowerment Zones, 182
Unwin, Raymond, 50
uranium mining, 32–34
urban development, 1–21
  EU policy, 9–10, 77
  functional economic planning, 60–61
urban governance, 10–14, 181–82
URBAN II Community Initiative Programme (2000-2006), 187–94
urban management, 6–14
  'capacity' approaches, 9
  'utilitarian' approaches, 8
URBAN programme (EU), 11, 182
urban regeneration, public participation in, 197–210
Urban Renaissance Report (UK), 77

Waterford city
  population, 2
  regional gateway, 16
Westmeath, population changes in county, 145
Wicklow, population changes in county, 145, 146